OTHER TITLES OF INTEREST FROM ST. LUCIE PRESS

Basics of Supply Chain Management

Applying TOC to Supply Chain Management

Total Productivity Management: A Systemic and Quantitative Approach to Compete in Quality, Price, and Time

Step-by-Step QFD: Customer-Driven Product Design

TRIZ Primer

Introduction to the Theory of Constraints

The Constraints Management Handbook

Integrating the Theory of Constraints and TQM: A Thinking Tools Approach

The Theory of Constraints Thinking Processes

Multi-Change Management

Fundamentals of Industrial Quality Control

For more information about these titles call, fax or write:

St. Lucie Press
2000 Corporate Blvd., N.W.
Boca Raton, FL 33431-9868

TEL (561) 994-0555 • (800) 272-7737
FAX (800) 374-3401
E-MAIL information@slpress.com
WEB SITE http://www.slpress.com

$S{^t_L}$

Supply Chain

MANAGEMENT

The Basics and Beyond

Supply Chain
MANAGEMENT
The Basics and Beyond

William C. Copacino

The St. Lucie Press/APICS Series on Resource Management

$S{^t_L}$

St. Lucie Press
Boca Raton, Florida

The Educational Society for Resource Management

APICS
Falls Church, Virginia

Phone: (561) 994-0555
E-mail: information@slpress.com
Web site: http://www.slpress.com

S^t_L

St. Lucie Press   APICS
2000 Corporate Blvd., N.W.   500 West Annandale Road
Boca Raton, FL 33431-9868   Falls Church, Virginia 22046-4274

DEDICATION

To Jan,

who has provided unending support to me,
while providing a loving home for our family
and care for her patients.

W.C.C.

TABLE OF CONTENTS

ACKNOWLEDGMENTS

This book organizes almost ten years of my columns from *Logistics Management* (formerly *Traffic Management*) into seven chapters around key themes in logistics and supply chain management. There are numerous people I want to thank for their assistance and support with my columns over the past decade and with this book. I first want to thank my clients for providing me the opportunity to observe and understand the emerging issues they faced and how they creatively responded to these issues. When I first began writing my columns, I had only a handful of ideas and I worried that I would run out of things to write about in less than a year. Fortunately, the experience of working with these creative and progressive clients has allowed access to a wealth of new ideas and innovative thought, and I thank these clients for this privilege.

I also want to thank *Logistics Management* for their support over the past decade. Past editor Frank Quinn took a chance and put his faith in me, based only on my limited vision for the column. Mitch MacDonald, his successor, has been equally supportive, and Toby Gooley has consistently provided outstanding editorial support.

I also want to acknowledge the guidance and encouragement provided by George Gecowitz of the Council of Logistics Management, Doug Lambert of the University of North Florida, Bud LaLonde of Ohio State University, and Jonathan Byrnes of Jonathan Byrnes and Company and the Massachusetts Institute of Technology. George reinforced the principles of integrity and high quality in everything one does in logistics, and Doug and Bud were always available to "think with," bounce ideas off, and enhance my thought process. Jonathan has been most helpful in the development of new ideas and in framing logistics and supply chain issues in the strategic context.

In addition, I want to thank my colleagues at Andersen Consulting, as well as my former colleagues at Arthur D. Little, Inc., for their ideas, constructive criticisms, and encouragement. In particular, Joe Martha and the late Ken Ernst joined me to initiate and build the Logistics Strategy Practice at Andersen Consulting, and the new directions we collectively developed for our clients provided valuable ideas for my columns. Also, Frank Britt, Larry Lapide, Don Rosenfield, Jamie Hintlian, Chris Lange, Steve Sotzing, Ken Mifflin, Paul Matthews, Doug Bade, Kevin O'Laughlin, Jack Barry, Vic Orler, Greg Owens, and Dave Andersen were valuable sources of new ideas. In addition, I thank Peter Fuchs, Joel Friedman, and Joe Forehand for providing an environment which fosters thought leadership, innovation, and new thinking.

Susan Gurewitsch has provided outstanding editorial support for this book, and Brian Kardon of Cahners Publishing has been supportive of and encouraging in developing these columns into a collection.

I was also fortunate to work with very talented and understanding executive assistants, who helped keep me organized and were always there to ensure I met all deadlines. Specifically, Roberta Green, Mona K. Altschuler, and Nancy E. Johnson have been invaluable partners and collaborators.

ABOUT THE AUTHOR

Willliam C. Copacino is the Managing Partner of Andersen Consulting's Strategic Services Practice in the Northeast. He has spent the past 19 years as a consultant to leading-edge companies in the areas of supply chain management, marketing strategy, manufacturing and operations strategy, and logistics strategy. He is broadly recognized as one of the leading thinkers and consultants in the areas of supply chain management. His clients have spanned many industry segments, including consumer packaged goods, retailing, pharmaceuticals, transportation, industrial products, electronics, and automotive companies.

Mr. Copacino has written extensively on supply chain management issues. He is the co-editor of *The Logistics Handbook* (The Free Press, 1994) and the co-author of *Modern Logistics Management* (John Wiley, 1985). He has also contributed to several other books, including *Reconfiguring European Logistics* (CLM, 1993), *The Change Management Handbook* (Irwin Professional Publishing, 1994), and *Reinventing the Warehouse* (The Free Press, 1993). For the past ten years, he has written a column on Logistics Strategy for Logistics Management magazine.

Mr. Copacino has an MBA from the Harvard Business School and a B.S. in Industrial Engineering and Operations Research from Cornell University. His wife, Dr. Janet Hall, and their three children, Michael, Steve, and Caroline, reside in Newton, Massachusetts.

ABOUT APICS

APICS, The Educational Society for Resource Management, is an international, not-for-profit organization offering a full range of programs and materials focusing on individual and organizational education, standards of excellence, and integrated resource management topics. These resources, developed under the direction of integrated resource management experts, are available at local, regional, and national levels. Since 1957, hundreds of thousands of professionals have relied on APICS as a source for educational products and services.

- **APICS Certification Programs**—APICS offers two internationally recognized certification programs, Certified in Production and Inventory Management (CPIM) and Certified in Integrated Resource Management (CIRM), known around the world as standards of professional competence in business and manufacturing.
- *APICS Educational Materials Catalog*—This catalog contains books, courseware, proceedings, reprints, training materials, and videos developed by industry experts and available to members at a discount.
- *APICS—The Performance Advantage*—This monthly, four-color magazine addresses the educational and resource management needs of manufacturing professionals.
- *APICS Business Outlook Index*—Designed to take economic analysis a step beyond current surveys, the index is a monthly manufacturing-based survey report based on confidential

production, sales, and inventory data from APICS-related companies.

- **Chapters**—APICS' more than 270 chapters provide leadership, learning, and networking opportunities at the local level.
- **Educational Opportunities**—Held around the country, APICS' International Conference and Exhibition, workshops, and symposia offer you numerous opportunities to learn from your peers and management experts.
- **Employment Referral Program**—A cost-effective way to reach a targeted network of resource management professionals, this program pairs qualified job candidates with interested companies.
- **SIGs**—These member groups develop specialized educational programs and resources for seven specific industry and interest areas.
- **Web Site**—The APICS web site at http://www.apics.org enables you to explore the wide range of information available on APICS' membership, certification, and educational offerings.
- **Member Services**—Members enjoy a dedicated inquiry service, insurance, a retirement plan, and more.

For more information on APICS programs, services, or membership, call APICS Customer Service at (800) 444-2742 or (703) 237-8344 or visit http://www.apics.org on the World Wide Web.

1 INTRODUCTION AND OVERVIEW

Ten years ago I began writing a monthly column on "Logistics Strategy" for *Transportation Management* because I saw logistics growing ever more strategic in focus and this fundamental shift going largely ignored. Few were looking at logistics from this new perspective; instead, many continued to think in more functional terms, i.e., logistics as transportation and distribution. Others thought in more operational terms, i.e., logistics as a lever to reduce costs or improve customer service. Some were beginning to think in supply chain terms—sales, logistics, and manufacturing as a linked system. Only a handful of people were thinking about the extended supply chain, suppliers-manufacturers-distributors-retailers-customers, as an integrated economic and operating system. Most notably, logistics was not linked to strategy. It was not considered as a critical component of the business, and it was rarely viewed as a source of competitive differentiation. I viewed these narrow perspectives as a mistake.

My vision in 1986 was that logistics and supply chain were emerging as more important and, in some cases, critical strategic variables. I have been writing about the strategic aspects of logistics and supply chain ever since. Time and fate have validated this view. Over the decade, logistics has assumed greater and greater importance as a strategic variable, so that today excellent logistics and supply chain management are essential elements of successful companies. Achieving this excellence requires thinking about logistics in terms of customers, competitors, costs, and connections with other activities across the entire extended supply chain.

The columns that fill this book look at supply chain management and logistics from all these perspectives. The selections span my decade as a columnist. I have not fundamentally changed any of the columns (but I have changed selected column titles), and each column includes its date of publication so the reader can understand the context in which it was written. Most of the developments discussed in even very early columns remain true today; thinking in some areas has advanced further, and some trends noted are just beginning to assume real importance.

I have organized the columns into six chapters that focus on key aspects of supply chain management and logistics:

- **Chapter 2, "Overview of Supply Chain Management and Logistics Strategy,"** defines supply chain and logistics management and related terms, outlines the requirements for excellence in these activities, and argues that integration (across functions, processes, and even organizations) is critical to success. Selected columns also clarify the differences between logistics planning and logistics strategy and provide a foundation for understanding logistics in the strategic context as well as an approach for developing a logistics and supply chain strategy.

- **Chapter 3, "Supply Chain Management and Logistics in a Competitive Context,"** takes a practical approach to the issue of how to use logistics to win and sustain competitive advantage, including how to organize and manage for success.

- **Chapter 4, "Customer Service,"** argues that excellent supply chain and logistics management begins with understanding customer needs and explores how to gain the requisite understanding and then how to translate it into superior capabilities.

- **Chapter 5, "Functional Excellence,"** views logistics as a process that integrates a number of activities—forecasting and inventory control; manufacturing; transportation and warehousing; purchasing; information management; and sales, marketing, and customer service—which must be executed excellently, individually and in combination, to make supply chain management effective.

- **Chapter 6, "Techniques for Supply Chain Excellence,"** outlines some practical tools and approaches for analyzing, and thus better understanding, key processes and costs of the supply chain.
- **Chapter 7, "Future Trends and Issues: The Broader Context,"** tries to avoid crystal-ball predictions of the future, but lays out some emerging directions to give management insights into key issues that are sure to have impact of some kind on effective supply chain management and logistics.

While these columns range widely, the collection is no encyclopedia of supply chain management and logistics. Mindful of the executive reader, each column can be read and digested in a few minutes. The goal is to highlight key issues, not to explore them in depth. My objective in writing these columns has been to raise awareness of logistics and supply chain issues in the strategic and competitive contexts and to provide some insight in how to analyze and address those issues. The limits of space (a single page in *Logistics Management*) and the need to tailor the approaches to the particulars of varied industries and specific companies preempted a comprehensive treatment of each area. Taken together, this collection reflects the progression of thinking in the field (and the mind of one consultant) over the past decade and provides a foundational understanding of the key concepts, issues, and trends in supply chain management.

I encourage the readers of this book to take these columns for what they are—bite-sized hors d'oeuvres which encourage you to consume another, rather than a full meal which will satisfy all your needs. However, collectively they provide an informed and digestible overview of the dynamic world of logistics and supply chain management. In that context, I hope this book provides valuable reading for experienced and less experienced managers, academics, consultants, and students of logistics and supply chain management.

2 OVERVIEW OF SUPPLY CHAIN MANAGEMENT AND LOGISTICS STRATEGY

The columns in this chapter amount to a definition of supply chain and logistics management. A key word that appears throughout these selections is *integration*. Its prevalence is no coincidence because integration spells the difference between the old view of logistics as the discrete functions of transportation and distribution and the new vision of supply chain management that links all the players and activities involved in converting raw materials into products and delivering those products to consumers at the right time and at the right place in the most efficient manner.

The 17 columns that comprise this chapter fall into four clusters:

■ The first eight selections provide definitions of the terms and the art and science of supply chain management and logistics, with particular attention to the challenges involved in excellent execution. Some of these challenges are easy to address; many are not. Included are my first two columns, "Integrated Logistics" (2.2—December 1986) and "Clarifying Terms" (2.1—February 1989), which provide a basic definition of terms. Also included is a more recent column on the "Seven Principles of Supply Chain Management" (2.7—January 1996).

■ The next three columns (2.9 to 2.11) position logistics within the extended value chain (suppliers-manufacturers-customers) and explore the resulting need to think outside the box of corporate boundaries to achieve success. While this notion is well accepted today, when I wrote my first article on this topic in 1983 and my first column in this area, "Taking Logistics Beyond Corporate Boundaries" (2.10—February 1987), few people embraced this concept. The column "Integration: The Next Phase" (2.9—September 1991) reflects a more mature view of the extended supply chain concept, parallels my introductory address to the 1991 Council of Logistics Management Conference, and reflects the conference theme.

■ Four columns (2.12 to 2.15) then examine the process of developing supply chain management/logistics strategies and draw important distinctions between strategy and planning. "Logistics Strategy vs. Logistics Planning" (2.13—December 1992) and "What Is Logistics Strategy?" (2.12—August 1987) articulate this distinction.

■ The last two selections (2.16 to 2.17) confront the need for logistics managers to establish the value of an effective logistics function in the minds of top management, few of whom have extensive experience and knowledge in logistics and supply chain management. Each of these columns, "Promoting the Logistics Function" (2.16—October 1989) and "Moving Beyond 'Just Say Yes' Logistics" (2.17—September 1993), provides advice on positioning the logistics function and supply chain concepts in your company.

DEFINITIONS

2.1 Clarifying Terms

February 1989

The logistics concept, defined as the integrated management of the forecasting, inventory-control, transportation, warehousing, order-entry and customer service, and production planning functions, is taking hold and is now broadly accepted in a growing number of companies. Even the term "logistics," once generally associated with materials management in the military, is becoming more commonly used throughout business. With this advance, as demonstrated by the Council of Logistics Management's relatively recent renaming, it is perhaps useful to clarify some terms, since they are often used inappropriately. Definitions are as follow:

- **Logistics** and **supply chain management** refer to the art of managing the flow of materials and products from source to user. The logistics system includes the total flow of materials, from the acquisition of raw materials to delivery of finished products to the ultimate users (as well as the related counterflows of information that both control and record material movement). As such, it includes the activities of sourcing and purchasing; conversion (manufacturing) including capacity planning, technology solution, operations management, production scheduling, and materials planning (MRP II); distribution planning and management industry warehouse operations; inventory management and inbound and outbound transportation; and the linkage with the customer service, sales, promotion, and marketing activities. In practice, many professionals use the term logistics to refer to the supply chain activities of transportation, warehousing, and finished goods inventory management. I use it here in the broader context of the full supply chain activities.
- **Physical distribution** refers to that portion of a logistics system concerned with the outward movement of products from the seller to customer or consumer.
- **Physical supply** refers to the portion of a logistics system concerned with inward movement of materials or products from sources or the suppliers.
- **Manufacturing planning and control** refers to the management of materials through a manufacturing facility and generally involves raw-material inventory control, capacity plan-

ning, production scheduling, shop-floor control, work-in-process inventory control, and purchasing.

- **Distribution** refers to the combination of activities and institutions associated with the advertising, sale, and physical transfer of products and services. It is concerned, therefore, with broader matters than logistics alone.
- **Extended supply chain** refers to the integrated set of activities completed by the full supply chain participants (suppliers, manufacturers, distributors, retailers/customers, and consumers/end users). It effectively includes the supply chain activities of each player in the channel.

With a common understanding of these terms, we in the logistics community should be in a better position to communicate our perspectives to our counterparts in other functions.

2.2	Integrated Logistics

December 1986

Many top managers continue to respond slowly to the facts that have been known by the logistics community for a long time—that integrated logistics management can improve *both* cost and customer service performance. Companies that have embraced "integrated logistics management," or the "total cost concept," have attained attractive competitive positions. The total cost concept of logistics is based on the interrelationship of supply, manufacturing, and distribution costs. Put another way, ordering, inventory, transportation, production setup, warehousing, customer service, and other logistics costs are interdependent. A change in any one of these activities influences the others, and an attempt to minimize any individual cost element may result in higher total logistics cost.

For example, setting a limit on inventory levels in a company with a seasonal demand pattern may well result in lower inventory levels. But the additional costs incurred by manufacturing in adjusting production levels to match demand may more than offset these inventory savings. A level production strategy may well result in lower total costs. Similarly, regional stocking may permit dramatic reductions in transportation costs through increased shipment consolidation, as well as expanded sales through better delivery performance. These improvements may be accomplished with only moderate increases in inventory and warehousing costs. However, in an environment where different functional units manage the various logistics activities independently, a company is less likely to properly analyze such important trade-offs.

Many companies have made important strides toward integrated logistics management. Several recent surveys have documented this trend, noting the expanded responsibility and importance of the logistics manager. In one highly publicized example, a major food producer combined all of the logistics functions within a single organizational unit. Not only were transportation costs significantly reduced through better coordination of inbound and outbound shipments, but integrated production and distribution planning led to improved service performance *and* reduced inventories. Subsequently, the company differentiated itself in the marketplace based on the superior customer service it provided through its well-managed logistics system.

Through ignorance, tradition, or marketing-dominated decision making, many companies have ignored or only partially adopted integrated logistics management. In today's service-sensitive environment, these companies will not be able to defer opportunities to both reduce cost and improve customer service performance for long.

2.3   Achieving Functional Integration

April 1992

Most businesses throughout the world organize their people and manage their activities through "functional groupings"—sales, marketing, manufacturing, finance, distribution, and so forth. A primary goal of these functions is to develop efficient ways to prosecute their work. The people in these functions become experts and seek to achieve superior performance in their function—what I call "functional excellence"—as measured by higher sales, lower transportation costs, lower inventories, or better control of operations.

Increasingly, companies are finding that functional excellence does not equate to business excellence, which is achieved through superior business coordination. In fact, two leaders in the field of organizational design, Paul Lawrence and Jay Lorsch, developed a theory that I have seen applied with profound success. It reads as follows:

- In a stable environment—one characterized by demand certainty, low seasonality, longer product life cycles, and low competitive intensity—it is best to organize for functional excellence. That is, organize each activity as a separate entity and encourage it to operate as efficiently as possible.
- In a more dynamic environment—one characterized by demand uncertainty, significant seasonality, short product life cycles, or high competitive intensity—companies that organize for *functional integration* tend to outperform those that are organized for functional excellence.

Today, virtually all industries and most companies are facing a more dynamic environment—that is, greater uncertainty of demand, shorter product life cycles, and so forth. Therefore, companies that organize for functional integration will almost certainly outperform those that organize for functional excellence.

There is no "silver bullet" answer for ways to achieve functional integration. Rather, one must address all aspects of a company's operations, as follows:

- Manage the process, not the function. Most major business processes (such as order fulfillment) cross functional lines. Many companies traditionally have focused on increasing functional efficiency—that is, addressing how they could improve each step in the process. A holistic view of the process invariably reveals that many of those steps are not necessary, that others are redundant, and that much time and money are wasted as work flows from one step to another. Therefore, more companies today are redesigning the core business processes with profound results.

■ Align measurement systems and incentives with overall goals. Cost per ton mile is a frequent measure of the transportation function's effectiveness. The application of this simple measure, however, can lead to suboptimal company performance. Measures and incentives must align with process management and overall corporate goals. Furthermore, any measures and incentives must not be the default for thoughtful, involved management. Management must temper measures with a thoughtful business perspective.

■ Utilize integrating mechanisms such as the sales and operations planning meeting (weekly or monthly), cross-functional teams, and team problem-solving approaches.

■ Work to develop a culture that encourages teaming and cross-functional collaboration. This can be accomplished through a variety of initiatives including mission/value statements, recognition of teaming efforts, and designing career paths to involve multi-functional assignments.

In the future, business success will be increasingly dependent on functional integration. Because logistics, by definition, is an integrating function, it is in a position to play a unique role in the integrated organization.

2.4   Managing the Pipeline

December 1989

An increasing number of companies are beginning to look at their operations in terms of a *pipeline* that manages the flow of materials

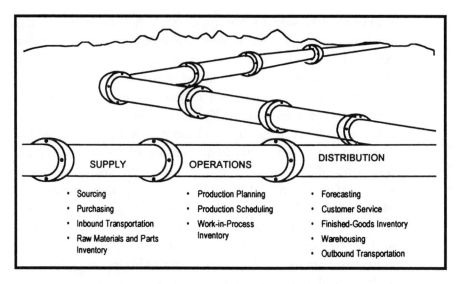

Figure 2.1   Integrated Pipeline

from source to ultimate consumer. Though not entirely new in substance (a similar framework, the total cost concept, has been discussed for a long time), the pipeline idea is proving to be an analytical concept that is able to transcend internal political obstacles and help achieve a quantum leap in functional integration and operational effectiveness.

The concept is simple and requires managers to think about their supply, operations, and distribution activities as an integrated pipeline (see Figure 2.1). The idea is to analyze the interaction of each of these activities as one integrated system and assess the performance of that system along the following three dimensions:

- **Cost**—The full cost of processing and moving materials from source to point of use.
- **Service**—Including issues such as delivery reliability, in-stock performance, and delivery lead time.
- **Velocity**—The time it takes to move products through the

logistics pipeline. This performance dimension is directly correlated to total pipeline inventory levels and to the flexibility of the pipeline to respond to changes in the market.

This pipeline approach provides a mechanism for analyzing the impact of each function's operating policies and for identifying conflicts and inconsistencies. For example, this full-system view will quickly identify the cost, service, and velocity impacts of any poorly conceived practices (such as the tendency among buyers to procure large order quantities and generate bloated inventories because they are measured solely on purchased price). In addition, it will provide managers with an efficient means to determine the following:

- The effect of long production runs and manufacturing lead times on system inventory
- The impact of marketing promotions on operating costs and on effective net margins
- The cost of or savings available from transportation consolidations
- The cost of multiple handling through a multi-echelon warehouse network

Most encouraging is the realization that the logistics pipeline concept has provided companies with a mechanism to achieve functional integration. This simple, elegant construct identifies the full performance impacts of all functional policies—be they manufacturing, marketing, finance, or distribution—in a way that is easy to comprehend.

The pipeline analysis, moreover, provides a compelling reason to modify narrow functional initiatives that do not add value to the business. The potential benefits of lower costs, better service, and increased velocity from a pipeline-oriented approach are visible and therefore build a case for action. Thus, the pipeline concept provides a way to move beyond internal political obstacles and to construct an enhanced operating capability.

2.5   Growing Focus on Inbound Logistics

May 1989

An important shift in the focus of the logistics function is occurring. For a long time now, the primary focus of many logistics organizations has been on the outbound side—the distribution function. The challenge has traditionally been for distribution managers to coordinate the functions of forecasting, finished goods inventory management, warehousing, transportation, customer service, and production scheduling. This often was a daunting task for managers in companies where these functions were spread across a number of departments, such as marketing, manufacturing, and even finance.

An even greater challenge faces managers today. The coordination of purchasing, inbound transportation, raw materials, component parts, work-in-process inventories, and production scheduling must be of increasing concern to logistics managers. This shift toward inbound management is the result of many factors, including:

- Transportation deregulation, which has encouraged shippers to coordinate the purchase of *all* transportation in order to minimize empty backhaul miles and maximize purchasing leverage
- The proliferation of just-in-time systems, which require close coordination of inventories with purchasing and delivery timetables
- Increased outsourcing, which extends the walls of the factory outward to include selected suppliers, but which also requires careful management of inbound shipments
- Enhanced technological capabilities, including information systems (MRP II) and communications capabilities (EDI)

The logistics managers of tomorrow will be pipeline managers. They will manage a complex flow of materials from a geographically diverse supply base that will feed global manufacturing facilities operating with minimum inventory. Companies will place more and more emphasis on maintaining the minimum assets, achieving low operating costs, and providing flexibility and superior customer service.

This shift in focus from primarily outbound traffic to both outbound *and* inbound traffic requires logistics managers to broaden their knowledge and capabilities. Managers will need an understanding of the entire purchasing function, including sourcing, specification, negotiation, procurement, and supplier performance measurement; an understanding of the supply chain, including even a knowledge of cost accounting and finance, to be able to identify leverage points and areas of opportunity; and a knowledge of information technologies, which have become key for managing these complex flows.

2.6 What Is "Fluid Distribution"?

April 1993

A handful of companies are aggressively moving to a new operating concept known as *Fluid Production and Distribution Operations*. This approach represents the ultimate form of supply chain management and channel integration, and it provides these companies with unmatched competitive capabilities.

One example of a company that has made considerable progress in developing a Fluid Production and Distribution Operation is Sony Corp. Sony calls its system *SOMO*—Sell-One-Make-One. Here are some of the key elements of this approach:

- **Inventory visibility**—Fluid Operations requires a view of product usage rates throughout the supply pipeline. Companies therefore use point-of-sale information to monitor usage rates at the retail level, to assist in long-term forecasting (note that short-term forecasting is eliminated under this approach), to measure promotional effectiveness, and to provide a view of inventory levels at all points in the supply chain.

■ *Manage flow, not replenishment*—Inventory management in a Fluid Operations system focuses on managing the flow of products through the channel, as opposed to traditional replenishment of depleted inventory. The Fluid Operations system requires that the manufacturer view the supply chain as a pipeline. The company therefore must focus on sizing the pipe correctly, as well as on managing the flow through that pipeline. The result is a "one out/one in" approach. This requires a fundamental rethinking of traditional inventory management techniques.

■ *Flexible distribution*—Distribution operations are set up for speed and flexibility. Therefore, cross-dock operations and plant-direct shipments play a key role in Fluid Operations, and faster modes of transportation often are utilized. Furthermore, this approach places a greater emphasis on inventory management than do traditional approaches to distribution.

■ *JIT manufacturing*—The Fluid Operations approach requires manufacturing to make quick changes, resulting in shorter, more frequent production runs. Furthermore, manufacturing must be flexible enough to respond to surges in demand, so flexibility normally is built into a Fluid Operations system through capacity-planning guidelines that support peak requirements.

■ *Interfunctional cohesion*—Functional roles are secondary to the management of processes in Fluid Operations. Measurement systems are designed to monitor total systems performance (total costs, aggregate asset performance, customer service, etc.), rather than function-specific measures (for example, transportation costs and manufacturing asset integration).

■ *Advanced information systems*—Advanced information systems are an integral part of a Fluid Production system. Visibility of current, accurate inventory status and product "flow rates," leading-edge application of software, and advanced decision-support systems all are essential components of Fluid Operations.

A Fluid Operations system provides companies with a competitive advantage by furnishing an operating profile that is low cost, flexible, and responsive to market needs and direction. The challenge in implementing Fluid Operations is that a company cannot go half way in building this capability. That is, most of the components described above must be in place before significant benefits are realized. Once in place, however, a Fluid Production and Distribution system provides a distinctive operating advantage.

2.7 Seven Principles of Supply Chain Management

January 1996

U.S. corporations have done nearly all that can be done to "optimize" operations within the four walls of their enterprises. Now they must look across the supply chain for the next step in improving the effectiveness of their logistics operations. Their objective is simple: to boost competitiveness and profitability.

The great benefit of supply chain management is that when all of the channel members—including suppliers, manufacturers, distributors, and customers—behave as if they are part of the same company, they can enhance performance significantly across the board.

Of course, this is easier said than done. Successful supply chain management is extremely complex; the number of players involved means that each company's supply chain management process will be unique. There are, however, some general principles that every company should follow when managing across the supply chain. My colleague, Jamie Hintlian, recently outlined seven of these principles. By working aggressively to realize these seven principles, your company can build a foundation for long-term competitive advantage.

1. ***Begin with the customer*** by understanding the customer's values and requirements. Grouping customers based on fulfillment requirements (such as delivery terms, merchandising support, and value-added services) vs. traditional categories (e.g., by class of trade) identifies how to align your operations to meet your customers' requirements.

2. ***Manage logistics assets*** across the supply chain, not just within the enterprise. Projects that address the location of distribution facilities, pipeline inventory, and transportation operations should include both down-channel and up-channel partners. For example, vendor-managed inventory programs will require collaboration to decide how information will be shared, where inventory should be deployed, and what appropriate performance criteria are.

3. ***Organize customer management*** so that it provides one "face" to the customer for information and customer service. This means aligning suppliers' fulfillment processes with your customers' buying processes. It also requires that information technology be leveraged to provide a single window on order status as well as electronic connectivity between supply chain members.

4. ***Integrate sales and operations planning*** as the basis for a more responsive supply chain. For example, both functions should have a single forecast number. This requires sharing real-time demand and forecast information both within the enterprise and across the supply chain.

5. ***Leverage manufacturing and sourcing*** for flexible and efficient operations. Leading companies are using other tactics in addition to just-in-time and "lean manufacturing" practices. Postponement strategies, for example, leverage manufacturing flexibility to reduce inventory while providing high service levels for multiple stock-keeping units. Automatic replenishment programs can be linked to production planning and scheduling. Companies also can use demand signals such as point-of-sale data as a barometer for marketplace activity. They should, in fact, link them to all planning processes.

6. ***Focus on strategic alliances and relationship management*** across channel partners. Although it is challenging to develop true partnership relationships, the fact is that without strategic partnerships, it is impossible to manage the supply chain as a single entity.

7. ***Develop customer-driven performance measures.*** Such measures ultimately drive the behavior of all channel members. The complete supply chain solution requires developing measures and performance criteria that track the economic performance of the extended supply chain.

2.8 | Integrated Supply–Demand Chain Management

June 1996

The consumer products industry has collaborated on a distribution strategy initiative called Efficient Consumer Response (ECR), which is designed to integrate and rationalize product assortment, product promotion, new product development, and product replenishment across the supply chain. (See Column 7.11 in Chapter 7 for a discussion of ECR.)

ECR's proponents believe it can enhance the efficiency of the $600 billion consumer products industry by more than $30 billion. Because ECR appears to hold so much promise, other industries have launched similar initiatives. The most recent example is the medical products industry's "Efficient Patient Response" program.

Though such efforts are valuable—they have, in fact, produced some substantial results—it is increasingly evident that they are insufficient. To explain why, I will take a look at ECR and its limitations and then suggest a much more powerful and complete approach that could be called "ECR Plus."

As I noted earlier, the ECR initiative has produced numerous benefits. It has grabbed the consumer products industry's attention and raised industry-wide consciousness of the huge and growing problem of non-value-added costs in the supply chain. ECR also has spurred cooperative efforts that have produced standards in such key areas as electronic data interchange, direct store delivery, cross-dock operations, continuous replenishment, and computer-assisted ordering. These efforts have led to the creation of some valuable resources, including value-chain analyses, a primer on activity-based costing, and a technology "road map."

Because most companies have been slow to act on any of these fronts, ECR has not yet achieved its full promise. Yet expectations about the industry's ability to implement ECR may have been unrealistic from the start. For example, inefficient invoice deduction processing and reconciliation continue to cost consumer products manufacturers and distributors billions of dollars each year.

Furthermore, the true promise of ECR is not clear. ECR, as it stands today, is a *vision* of efficiency, not a program for achieving it. That vision now focuses primarily on the supply side of the value chain. To succeed however, ECR must be incorporated into a broader perspective on building economic value.

This new perspective informs a process model that integrates supply chain management (the focus of ECR) with demand chain management. Demand chain management requires collaboration between manufacturers and retailers to build demand via effective consumer marketing and merchandising. Thus, only efforts to increase the efficiency of processes and operations, coordinated with efforts to build consumer demand, can realize the full potential of ECR.

Achieving success with ECR will require a truly integrated approach to supply and demand chain management—an approach that delivers what consumers want, where and when they want it, as efficiently as possible. I call this approach "Integrated Supply–Demand Chain Management," and it is illustrated in Figure 2.2.

What does this approach mean in practice? For one manufacturer, it meant segmenting customers according to the customers' merchandising strategies and operating sophistication and then tailoring the manufacturer's marketing program to each customer segment. This manufacturer developed coupons of differing values for

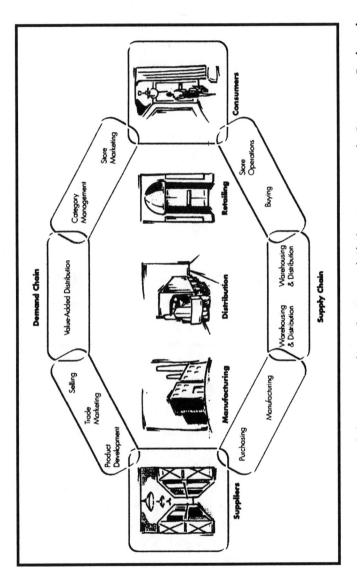

Figure 2.2 Framework for Integrated Supply–Demand Chain Management in Consumer Packaged Goods

retailers to use to capture information on low, medium, and heavy consumers of the manufacturer's product. It also offered customized promotions that would fit retailers' individual strategies and boost sales. For key accounts, the manufacturer also developed account-specific customer service measures, assigning multi-functional account teams that integrate operations and marketing planning for those customers. Finally, by using scanned point-of-sale data, the manufacturer enhanced store-level planning and merchandising for key retail accounts.

This focus on effectiveness complements the efficiencies of ECR and positions companies to develop an integrated and powerful competitive model. I am confident that the next wave of logistics excellence will be realized through Integrated Supply–Demand Chain Management.

LINKS WITH THE EXTENDED VALUE CHAIN

2.9	Integration: The Next Phase

September 1991

The Council of Logistics Management has announced that the theme of its 1991 conference will be "Integration: The Next Phase," an appropriate subject given the historical evolution of the field of logistics. A look at that evolution can be helpful in understanding the challenges and opportunities facing the field today.

Many could argue that the concept of integrated logistics is an old theme, dating back to 1916, when Arch W. Shaw's textbook, *An Approach to Business Problems* (Harvard University Press), first discussed the strategic aspects of logistics. A decade later, Ralph Borsodi's *The Distribution Age* (D. Appleton & Co., 1927) became the first text to define the term "logistics" as it is used today. But it was not until 1956 that the concept of total cost analysis was applied to the logistics field in Howard Lewis's article "The Role of

Air Freight in Physical Distribution." And the term "integrated logistics" was not coined until the early 1960s in writings by Bud LaLonde and Don Bowersox.

Although an old concept, integrated logistics has taken on new meaning in the 1990s. Bud LaLonde has called logistics a "boundary-spanning activity." I like to call it an integrating activity and find it useful to think of integration in two ways—internal integration and external integration.

Internal integration encompasses the traditional thinking embodied in the total cost concept and involves the coordinated management of a company's operational activities. The basic premise is simple: There are trade-offs among the different components of logistics (warehousing, transportation, inventory, customer service, purchasing, and manufacturing setup costs), and to achieve optimal performance, an organization must design and manage its logistics system in an integrated way. Although this thinking dates back to the early 1960s, only about one-third of the companies operating today have achieved internal integration of their logistics activities.

External integration refers to the integration of logistics activities across the supply chain. It has been given many names, including inter-corporate logistics, supply chain management, channel integration, quick response, and partnerships, but the basic features are the same: the coordinated management of logistics and deep operating linkages with suppliers, customers, carriers, and other third-party service providers.

External integration represents a new wave of thinking in logistics, and more and more companies are working aggressively to develop these capabilities. Nonetheless, fewer than 5 percent of companies today have developed comprehensive channel integration capabilities. Those that have, however, are discovering that this form of integration can both reduce costs and improve customer service significantly. In fact, external integration has changed the competitive dynamics in the industries where it has been applied.

The theme of "Integration: The Next Phase" is therefore timely and appropriate. Many companies are demonstrating renewed interest in achieving internal integration as they recognize the competitive importance of an effective logistics capability. At the same time, leading-edge companies are aggressively pursuing the second wave of integration—external integration—as a way to achieve a distinguished competitive advantage.

2.10 | Taking Logistics Beyond Corporate Boundaries

February 1987

A significant new trend has been evolving in logistics management in recent years—one that involves the collaboration of all participants in the supply chain in order to reduce the total logistics system costs. It has been referred to as "Supply Chain Management," "Logistics Partnerships," or "Inter-Corporate Logistics Management." For purposes of this discussion, I have labeled this trend the "Total Systems Approach" or TSA.

In the traditional logistics "total cost concept" model, companies worked to manage logistics as an entity and to lower the total logistics costs to the organization. The model involved balancing trade-offs among production run lengths, inventory, transportation, warehousing, and customer service.

Recently, however, an increasing number of companies have come to realize that though the total cost concept may be useful, it is limited because it does not consider the efficiency of the entire supply chain. TSA systems, on the other hand, involve the active collaboration of two or more participants in the supply channel (supplier, manufacturer, distributor, and/or customer) to manage all the logistics resources in the most efficient manner possible.

A few companies have already extended the TSA concept to include almost the full integration of logistics activities with their customers. One major food producer, for example, has established electronic information linkages with its customers in order to have current information on hand concerning sales rates, inventory levels, and customer service requirements for each of the customer's stocking locations. The food producer, moreover, actively manages the inventories at the customers' locations, both initiating and filling replenishment orders to meet agreed-upon customer service objectives.

This integrated logistics system provides the food producer with "advanced" information for forecasting, production planning, inventory planning, and the management of its distribution system. It also permits the fulfillment of replenishment orders with consideration given to the production and distribution economies of the entire supply chain.

TSA systems are not easy to establish, but as several companies have already found, the strategic and operational benefits are considerable. Fine-tuning one's own logistics system is no longer enough. The next quantum leap in logistics efficiency will come through the implementation of TSA systems, and those who are successful in these efforts will secure a formidable, and likely durable, competitive advantage.

2.11 Quick Response Gaining Favor

June 1990

The concept of "quick response" is gaining broad favor as companies in all parts of the supply chain develop an appreciation of its potent benefits. Those companies that have implemented such programs over the past five years have achieved powerful competitive advantages; those that have been slow to pursue quick response are racing to catch up.

Quick response involves the integration of the supply chain, effectively linking retailers, suppliers (manufacturers/distributors), and carriers in close communication and integrated decision making. Key elements of quick response include:

- ■ *Point-of-usage data capture*—Quick-response programs rely on the capture of item-level usage information at the point of sale or transfer. This generally involves technologies such as point-of sale scanning, UPC bar codes (or other bar codes in non-retail-sector industries), and bar codes on secondary container packaging.

- **Item-level management**—Powerful forecasting and inventory-control applications software is needed to maintain tight control over individual stock-keeping units that may number into the hundreds of thousands.
- **Rapid communication**—Successful quick-response programs depend on electronic data interchange (EDI) communication for most transaction sets (ordering, order acknowledgment, shipment advisories, debit memos, etc.) and among all parties (suppliers, carriers, retailers).
- **Partnerships**—Developing close operating relationships with key trading partners in the supply chain is essential for success. Quick response is easier to implement and works more effectively with a limited number of suppliers.
- **Discipline and commitment**—In order to work, quick-response systems require top management's commitment and leadership, broad training, and discipline and attention to detail.

Effective quick-response systems are not easy to achieve; companies must hurdle technical, human resource, and financial obstacles to develop and implement them. Nonetheless, rich rewards are available to companies that succeed. These benefits include lowering inventories by as much as 40 percent, improving in-stock availability significantly, cutting transaction and administrative costs in half, reducing replenishment lead times to a third or less of their former levels, identifying slow-selling items sooner, and reducing operating costs for all players in the supply chain.

Logistics and information systems managers must take the lead in developing a quick-response program. Companies that invest now to build a quick-response capability will be positioned to offer better customer service at lower costs tomorrow.

THE SUPPLY CHAIN PLANNING PROCESS

2.12 | What Is Logistics Strategy?

August 1987

The concept of logistics strategy is just now fully evolving. Like evolving ideas, logistics strategy has not attained a precise meaning, universally accepted and used by all managers. Thus, today, the term "logistics strategy" still means different things to different people.

I have found it particularly useful for top managers to distinguish between logistics strategy and logistics planning. This column offers a definition for each and presents an analytical framework—the cost–service trade-off curve—which is a useful tool for developing a logistics strategy as well as for evaluating alternative logistics plans.

First the definitions. Logistics strategy involves the determination of what performance criteria the logistics system must maintain—more specifically, the service levels and cost objectives the logistics system must meet. Because cost and service normally involve a trade-off, a company must consciously consider that trade-off and determine the desired logistics performance. This process involves consideration of the company's strategic objectives, its specific marketing strategy and customer service requirements, and its competitors' cost–service position.

Logistics planning, on the other hand, involves the deployment and management of all logistics resources in order to attain the desired cost–service performance. Considerations might include number and location of warehouses, type of warehouses, mode and carrier selection, inventory positioning, inventory levels, order-entry technologies and systems, and so forth.

The cost–service trade-off curve (Figure 2.3) is a useful device for clarifying the distinction between logistics strategy and logistics planning. The schematic's coordinates identify the cost and service

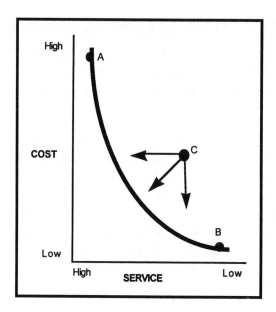

Figure 2.3   Cost–Service Trade-off Curve

options available to a company; the curve itself represents the optimal choices available.

For example, a company can select a high service position (Point A). This might imply more warehouses, use of premium transportation modes, and higher inventory levels so that high service levels can be provided (fast delivery, few stockouts). This high service position, of course, entails higher costs.

Alternatively, a company can select a low-cost position (Point B). This option might imply a single warehouse, longer order lead times to allow for more consolidation, and use of less costly transportation modes.

In reality, many companies' logistics operations are not at maximum efficiency, and they are located off the curve (Point C). These companies must decide where they want to be represented on the curve (the logistics strategy decision) and how to get there—that is, how they can utilize their logistics resources most efficiently to attain the desired logistics strategic position (the logistics planning decision).

In practice, these decisions must be made iteratively (that is, one affects the other). Nonetheless, I think there are many benefits from addressing the logistics strategy and logistics planning processes as separate but related decisions. I have found that too many logistics managers focus too heavily on the logistics planning decision. As a result, they do not devote sufficient time and thought to the development of a logistics strategy that is consistent with company objectives, customer service requirements, and competitive realities.

2.13 | Logistics Strategy vs. Logistics Planning

December 1992

After observing the logistics planning process at scores of companies, I find that some are better off if they distinguish logistics strategy development from logistics planning. Three benefits may accrue to companies that distinguish between these two activities:

- Opportunities for differentiation—based on operational, logistics, or customer service excellence—are more likely to be exploited.
- Logistics tends to have a more visible and more important role in the company.
- Investments in the logistics function or infrastructure are more likely to be approved.

In companies where logistics strategy and logistics planning are not distinguished from one another, the process tends to evolve into a budgeting exercise. These companies have more of a tendency to view logistics as a cost rather than as a value-adding activity.

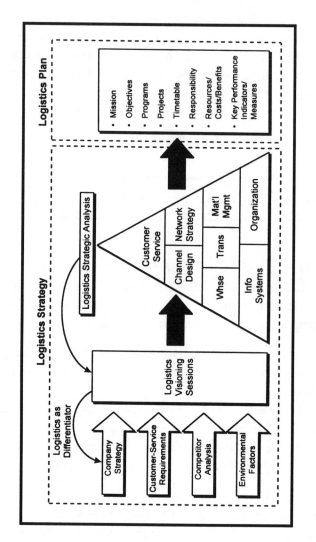

Figure 2.4 Logistics Strategy Development and Planning

I recommend the process depicted in Figure 2.4 for logistics strategy development and planning.

What follows is a brief discussion of each of the four elements:

- ■ *Critical inputs*—Key inputs to the logistics strategic planning process are an understanding of your company's strategy, a thorough understanding of customer service requirements, a view of competitor approaches, and insight into key external factors such as evolving regulatory requirements and the direction of freight rates.

- ■ *Visioning sessions*—Visioning sessions are meetings in which managers from key functional areas such as sales, marketing, and finance review the critical inputs and assess the implications for logistics in terms of cost, customer service, value-added capabilities, operating flexibility, and the ability to support innovation. Potential logistics strategies are explored. In particular, participants give explicit consideration to how logistics can be used to differentiate the company in the marketplace and/or support key strategic or customer service goals.

- ■ *Logistics strategic analysis*—This analysis involves the thorough consideration of all levels of the logistics pyramid—customer service strategy; structural considerations (such as channel design and network strategy); functional excellence in warehousing, transportation, and materials management; and implementation issues involving information systems capabilities and organization.

- ■ *Logistics planning*—Finally, the logistics plan is summarized, outlining objectives, programs, milestones, and key measures of performance.

Increasingly, companies are finding ways to differentiate themselves from competitors and position themselves favorably with customers through logistics. A thoughtful approach to logistics strategy development, including but extending beyond logistics planning, will be more important for companies in the years ahead.

2.14 | The Logistics Strategy Pyramid

September 1994

The term "logistics" has evolved to the point where it now has a commonly understood definition: management of the flow of product from source to point of use. But the concept of a logistics *strategy* and how it relates to all the components of logistics has remained ambiguous and often confusing for practitioners. If you ask ten logistics managers what key issues must be sorted out as part of the logistics strategy development process, you would likely get ten different answers! This lack of clarity sometimes has prevented logistics from becoming a key strategic consideration.

One useful framework for identifying and thinking through all the issues involved in developing a logistics strategy is the Logistics

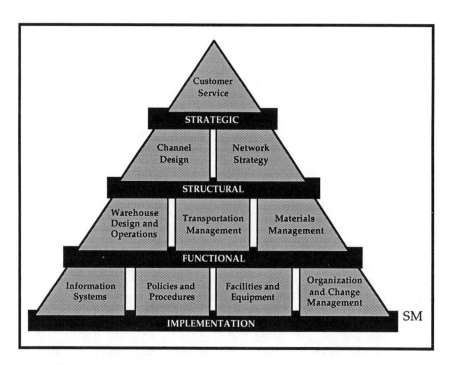

Figure 2.5   Logistics Strategy Pyramid

Strategy Pyramid. As shown in Figure 2.5, the pyramid allows issues to be analyzed on four levels:

- **Strategic**—On the strategic level, it is important to under-stand how logistics can contribute to your company's basic "value proposition" to your customers. Key questions to be addressed here include: What are the basic and distinctive service needs of our customers? What must logistics do to meet those needs? Can we use our logistics capabilities to provide unique services to our customers?

  Here's an example: In the mid-1980s, managers at medical products manufacturer Kendall Corp. recognized that they could add considerable value to their customers' operations by reducing the amount of inventory those customers held. Kendall established one of the first—and more successful—continuous replenishment programs using the "channel DRP" approach. In so doing, Kendall has achieved important com-petitive advantages.

- **Structural**—Once managers understand the value proposition and logistics strategy, they must address structural issues and ask questions such as: Should we serve the market directly or should we use distributors or other intermediaries to reach our customers? What should our logistics network look like? What products should be sourced from which manufacturing locations? How many warehouses should we have? Where should they be located, and what is the mission of each facility (full stocking, fast-moving items only, cross-docking, etc.)?

  These decisions are very complex and require careful analy-sis. However, given the innovative technologies and new operating options available to companies today, these struc-tural decisions provide opportunities for companies to *create value* and to achieve more for less. In other words, they can deliver enhanced, even innovative, customer service *and* re-duce their costs at the same time.

- **Functional**—On the next level, companies must ensure that they operate in the most effective way. Achieving functional excellence requires that they design optimal operating practices for transportation management, warehouse operations, and materials management (which includes forecasting, inventory management, production scheduling, and purchasing). Achieving functional excellence also entails developing a process-oriented perspective on replenishment and order fulfillment so that all the activities involved in these functions are well integrated.

- ■ *Implementation*—Without successful implementation, the development of logistics strategies and plans is meaningless. Of particular importance are organizational and information systems issues. Organization centers on the overall structure, individual roles and responsibilities, and measurement systems needed to build an integrated operation. Information systems are "enablers" for integrated logistics operations and therefore must be carefully designed to support the logistics strategy. Logistics managers must consider their information needs relative to decision-support tools, application software, data capture, and the system's overall structure.

It is important to note that decisions made within the Logistics Strategy Pyramid are interdependent. That is, you must understand what capabilities and limitations affect the functional and implementation decisions and consider those factors when developing a logistics strategy and structure.

The pyramid is an effective tool for identifying the complete set of questions that must be asked when developing a logistics strategy. It also provides a useful framework for integrating these decisions.

2.15   Using Visioning for Supply Chain Planning

May 1994

Intensified, global competition has placed increased pressure on almost all companies to improve their operations. What they must do, in effect, is increase productivity and lower their costs while at the same time providing greater customer value and customer service.

Companies that have not responded to these considerable pressures have at times experienced significant economic dislocations, including lost business, lower market share, and reduced margins. In some cases, they have had to exit from a market altogether. The market economy has been and will continue to be very harsh, and the market pressures for continued improvement will be relentless.

Many companies have not been very successful in facing these pressures because they have taken a narrow, functionally focused approach to operational improvement. Although those efforts have resulted in substantial improvements, they often have fallen short of restoring companies to their previous vibrant positions.

One reason why functionally focused initiatives don't work well is that solutions often involve changes to multiple functions, not just improvements in a single function. Another reason is that functionally focused initiatives address narrow concerns and pre-empt a broader examination of the issues. Typically, more radical, "out-of-the-box" thinking is necessary to address broader issues effectively. For example, a functionally focused approach would be to launch an initiative to reduce warehouse handling costs and enhance warehouse productivity. A broader approach to the same problem would be to ask whether the company could operate without a warehouse altogether.

Companies also have a hard time addressing the broader issues because their organizational structures force them to think in terms of narrow "slices." Logistics managers can add tremendous value to their companies by pushing management to think in cross-functional, boundary-spanning, integrated supply chain terms. Thinking this way can open new horizons for companies and encourage them to discover solutions well beyond their current, narrow perspective.

An effective way to develop that kind of thinking is called "supply chain visioning." Under this approach, the company holds a series of working sessions among top functional managers. The purpose of these meetings is to focus on developing a common understanding of the customer requirements in each segment of your business, of the unique "value proposition"—that is, the distinctive competence—that your company can use to better serve these customers' needs, and of how your operations must be changed to be brought in line with this vision.

Supply chain visioning is most successful when it focuses on customers, strategies, or results as follows:

- **Customer-focused**—This approach should be linked to a thorough understanding of customer needs. Often, a considerable amount of research is required in advance of the working sessions to deepen the understanding of customer requirements and of how competitive offerings fulfill those needs.

- **Strategy-driven**—Supply chain visioning must be consistent with the way your company has chosen to compete. It must support, for example, the unique value proposition that your company provides to customers. The working sessions offer an excellent opportunity to clarify and challenge your company's strategies and value proposition. At a minimum, companies benefit from this process because it ensures that their strategies are better understood and accepted and therefore are more effectively implemented by all functions.

- **Outcome-based**—The supply chain vision should focus on desired results. It is often useful to propose bold outcomes or to establish targets that require all involved to stretch their capabilities a little, thereby challenging participants to think in new ways. Some examples of such targets include cutting inventories in half, reducing cycle times by at least 60 percent, completely eliminating ordering and billing errors, or reducing total costs by 25 percent or more.

It is valuable to use "what if" scenarios to push reluctant participants to think in new ways. For example, a useful question might be: What would be the value to our company in the marketplace if we were able to provide same-day delivery (as opposed to our current ten-day delivery)? With a common sense of this market value, participants become bolder in their search for innovative ways to provide differentiating capabilities.

Supply chain visioning is an effective approach to forging functional integration, for pushing innovative thinking, and for linking operations to strategy. Furthermore, it provides a lever for logistics managers to fulfill one of their primary roles: enhancing cross-functional linkages and developing boundary-spanning solutions.

POSITIONING THE SUPPLY CHAIN AND LOGISTICS WITHIN YOUR COMPANY

2.16 Promoting the Logistics Function

October 1989

In most industries, logistics is becoming a more important business function. In many companies today, top management is giving broader recognition to the value created when shipments are complete, on time, and damage-free. Furthermore, as manufacturing costs decline, logistics costs are becoming more visible and more important.

Nonetheless, I believe that logistics managers need to do more to make top management aware of the value being added by an effective logistics function. Many logistics managers still understate their case or do not state their case at all. Given that many top managers come from finance or marketing backgrounds and do not have an intimate understanding of the functioning or the value of an effective logistics operation, this can be a serious tactical error.

I have visited countless companies where the addition of one logistics analyst could save hundreds of thousands of dollars, but the logistics departments have been unable to secure the additional resources. Similarly, I have seen numerous companies where improved software (order entry, transportation planning, or inventory management) could dramatically improve customer service and yield considerable savings, yet the logistics departments were unable to obtain approval to purchase the software.

I offer several suggestions for achieving higher visibility and recognition for the logistics function. They are as follows:

- **Speak from the customer's perspective**—My experience indicates that shippers consistently underestimate the importance of the service element in the marketing mix. Most shippers have opportunities to increase market share by improving their delivery and customer service performance. Yet market resources continue to be overallocated to advertising, promotion, and pricing discounts. Some simple customer research, focused on the service needs of customers in relation to other marketing variables, can highlight these opportunities.

- **Consider the revenue impact**—In making budget proposals, many logistics managers frame the benefit side of the analysis only in terms of cost savings. More attention should be given to the impact on revenues, that is, increased sales. In many cases, logic dictates that improving delivery performance or lowering costs has to lead to more sales; if stronger evidence is needed, however, there are well-proven analytical approaches for documenting the increased sales potential.

- **Frame the analysis in financial terms**—Make an attempt to translate the effects of any action to key financial considerations such as return on sales, return on equity, return on assets (ROA), or even to the impact on your company's stock price. For example, reductions in inventory have a leveraged effect on ROA because they affect both the numerator of the ROA equation (that is, they increase profits by lowering the costs of financing and storing the inventory) and the denominator (they lower the asset base through reduced inventories). One logistics manager obtained approval for a new order-entry and inventory-control system by presenting the following table as part of his analysis:

Increase in Net Return on Assets From Inventory Reduction

	Percent Improvement of ROA*				
Inventory as a Percentage of Total Investment	10%	20%	30%	40%	50%
10%	3%	5%	8%	11%	14%
20	5	11	17	23	29
30	8	17	26	36	47
40	11	23	36	51	67
50	14	29	47	67	89
60	17	36	59	84	114

* Assumes 25 percent inventory carrying cost and 15 percent initial ROA.

For his company, which had a substantial part (30 percent) of its assets tied up in inventory, he linked a 20 percent potential reduction in inventory levels from the new system to a 17 percent improvement in the company's ROA. Needless to say, top management approved funding for the order-entry system.

2.17 Moving Beyond "Just Say Yes" Logistics

September 1993

Most manufacturing companies today are being pressed by their customers to provide more for less—that is, lower prices, greater value, higher levels of customer service, and additional value-added services.

Clearly, many of these demands fall directly on the logistics department, with customers requiring more frequent deliveries, shorter order cycle times, higher in-stock availability, more customer pickup options with higher allowances, and increased drop shipments. Other popular requests include mixed pallets or links with cross-dock operations, expanded electronic data interchange capabilities, tighter appointment schedules, expanded responsibility for continuous replenishment programs, and various other value-added programs.

Many logistics managers find themselves with limited negotiating leverage relative to these "requests." They feel they have no real choice but to "just say yes" and meet these surging customer demands. At the same time, these logistics managers are being pressured by their own companies to enhance productivity and cut costs.

There are no clear and easy paths for logistics managers to avoid the "just say yes" trap. Clearly, the distinctiveness and brand strength of your company's products help to ameliorate these pressures. Beyond this obvious imperative, some companies have managed these more-for-less demands by bundling logistics activities with marketing programs and offering tailored solutions for their customers. In essence, they have positioned themselves in one of the four quadrants shown in Figure 2.6. Here's how the various alternatives stack up:

- Offering a "menu of logistics services" (lower left quadrant) is the least sophisticated approach, but it can be very helpful in preempting customer pressures to provide more for less. By presenting a menu of options, logistics managers can explicitly link these additional services both to the costs of providing these services and to the benefits and value to the customer. Although this approach will not always forestall customer demands for more services at lower cost, it does establish a framework where managers at least can consider the costs of providing the service and the benefits to the customer.

- Bundling the marketing and logistics services (top row) is a preferred approach for many companies. With this method, the costs of individual initiatives are less visible; therefore, it is harder for customers to demand additional services at no cost. The bundled marketing and logistics package forces customers to consider the value and benefits of the whole package.

	Static	Dynamic
Link With Marketing Programs	Menu of bundled service options with benefits and costs	Decision-support capability to assess cost and profitability impact for all supply-chain players of customized marketing and logistics program
Logistics Services Alone	Menu of logistics services with benefits and costs	Decision-support capability to assess cost and profitability impact for any combination of logistics program

Breadth of Programs

Analytic Capabilities

Figure 2.6   Alternative Approaches for Proactive Customer Management

- An automated decision-support system, which assesses the costs and benefits of customized programs (right-hand side of matrix), is a valuable tool for managers seeking to determine how far they can afford to go in responding to customer demands. This capability provides a way to rapidly assess the cost and profitability of any combination of logistics and marketing programs for a specific customer. Such a program will become an invaluable tool for many companies as customer demands accelerate.

As power shifts to the end of the supply chain in most industries, suppliers must innovate and develop distinctive products and services to avoid being managed as a commodity supplier by their customers. In addition to this imperative, suppliers can preempt and manage customers' demands that they provide more for less by developing and understanding the cost and benefits of individual marketing, sales, and logistics initiatives and by bundling these initiatives into integrated programs. The alternative is a continual margin squeeze and a less attractive business picture.

3 SUPPLY CHAIN MANAGEMENT AND LOGISTICS IN A COMPETITIVE CONTEXT

This chapter explores how to use supply chain management and logistics to achieve and sustain competitive advantage. The selections here focus on practicalities rather than theory—how to use logistics as a source of differentiation and how to organize and manage to capitalize on this possibility.

I believe that many of these columns reflect my best writing and my major mission, which was to frame the logistics issues and the logistics manager's mindset in strategic terms.

The 16 columns in this chapter fall into two clusters:

- The first eight selections explore specific ways that logistics can contribute to competitive advantage, ranging from creating an important cost advantage to customizing distribution to meet the needs of specific customer segments. The first two columns, "Creating Real Advantage from Logistics" (3.1—November 1992) and "Logistics Can Provide that Competitive Edge (3.2—June 1987), define the approach for understanding logistics in the competitive context, while "Delivering More for Less" (3.3—February 1991) creates a "sound-bite" which helped scores of logistics managers communicate the tremendous value available through effective logistics and supply chain management and assisted in marshaling resources to invest in achieving "more for less."

■ The other eight columns (3.9 to 3.16) discuss various structural approaches to achieving this competitive edge. Some focus principally on enhancing internal capabilities; others require establishing new relationships with channel partners. "The Perils of Partnerships" (3.13—February 1992) helps manufacturers and retailers to understand the difficulties, risks, challenges, rewards, and limits of partnerships.

GAINING STRATEGIC AND COMPETITIVE ADVANTAGE FROM LOGISTICS AND SUPPLY CHAIN MANAGEMENT

3.1	Creating Real Advantage from Logistics

November 1992

Much has been written about topics such as "Logistics as a Strategic Weapon," "Creating Competitive Advantage from Logistics," and "Value-Added Logistics." I find, however, that many of these writings fail to cut to the core and explain how companies can create *real* advantages from logistics.

I believe that there are five ways in which managers can use logistics to significantly improve their companies' competitive advantage. Specifically, they can offer:

■ *Low cost*—Through superior efficiency, logistics can contribute to a cost advantage that can be leveraged to increase market share or enhance profitability. Low-cost logistics is particularly important in logistics-intensive industries where there is little differentiation among commodity-like products (such as certain segments of the chemical or paper industries). In these industries, logistics costs can exceed 15 percent of the cost of goods sold.

- ■ ***Superior customer service***—The most notable measures for customer service include short order lead times and in-stock availability. Such measures also may include order and invoice accuracy, access to information on order status, ability to respond to customer inquiries, and so forth.
- ■ ***Value-added services***—This means providing services that enhance your customer's ability to compete and includes activities such as pricing and ticketing of merchandise, assembling mixed pallets, making drop shipments, delivering direct to stores, arranging for quick replenishment or continuous replenishment, and providing training or software to the customer.
- ■ ***Flexibility***—A logistics system can create an advantage by being flexible enough to customize its service and cost offerings to meet the needs of distinct customer segments or individual accounts. The flexibility to meet diverse customer needs in a cost-effective way can distinguish a company and allow it to serve a broader customer base.
- ■ ***Innovation***—Genuine value and competitive advantage can be provided by a logistics system with the capability to reinvent itself. By this, I mean it has the capability to innovate or develop new ways to serve a market. For example, because manufacturers of generic pharmaceuticals have begun to serve the channel direct, wholesalers of those pharmaceuticals have had to reinvent themselves to focus on high-service niches. Innovation requires an organization that has several important characteristics: it must have the capability to learn to change, flexible information systems that can adapt to new ways of doing business, the vision to recognize the need for and the direction of change, and the leadership to drive that change.

When thinking about how logistics operations can create value for their companies, logistics managers should explore these five "levers." Logistics capabilities do not necessarily have to be superior in all five of these areas. Generally, they must stand out in one or two of the five categories and be adequate in the remaining areas. This framework not only provides a structure managers can use to examine how the logistics function is adding value and how it is aligned with their companies' needs, but also represents a way to clarify and communicate their goals to others within their companies.

3.2 | Logistics Can Provide that Competitive Edge

June 1987

In recent years, companies in a variety of industries have focused increased attention on their basic operations capabilities. This focus has extended beyond just manufacturing and has included logistics and distribution activities. In fact, logistics itself has become an increasingly important basis of competition in many industries. Astute companies in a variety of fields have used logistics to differentiate themselves in the marketplace.

The emergence of logistics as an important strategic variable has occurred in many industries. Consider the changes that have taken place in the following sectors:

- **Consumer packaged goods**—Though the establishment and maintenance of a "brand franchise" remain important to consumer packaged goods manufacturers, their delivery and customer service capabilities are an increasingly important basis of competition. The abilities to provide on-time deliveries, offer value-added services, and reduce customers' costs through tailored distribution services are emerging as key success factors for consumer packaged goods companies.
- **Computers**—While product features and price remain critical competitive factors, logistics capabilities have grown in im-

portance. The abilities to process orders rapidly, control inventories (including regional stocks) effectively, and manage deliveries efficiently are necessary for computer manufacturers to be successful today.

■ *Medical supplies*—Price and delivery performance have joined product performance as the key bases of competition for medical suppliers. The suppliers with superior order-entry, information systems, and logistics capabilities are emerging as industry leaders.

Although the emergence of logistics as an important competitive element is perhaps most striking in these three industry sectors, it has occurred to varying extents in many other fields as well. This continuing trend places emphasis on logistics as a critical business function and creates new opportunities and challenges for logistics managers. Along with their complement of operating and planning skills, logistics managers will need to master new skills such as the ability to perform customer needs assessments, competitor evaluations, and strategic positioning analyses.

3.3   Delivering More for Less

February 1991

Logistics and transportation managers are being put in a difficult position. In company after company and industry after industry, they are being pressured to deliver "more for less"—improved customer service but at a lower cost. Customers are demanding heightened service performance—that is, shorter lead times, more frequent deliveries, elimination of stock-outs, and delivery within specified time windows—as well as value-added services such as bar coding of secondary containers, shrink wrapping of pallets, use of inner packs, and even direct store delivery. And these customers expect this heightened service at the same—or even lower—cost.

The pressure to improve both cost and service performance also is coming from the logistics manager's own employer. Sensitized to the heightened competitive importance of superior delivery performance, top management, too, is asking for more from logistics.

To meet these new demands, logistics and transportation managers cannot just fine-tune old ways of doing business. They must restructure the basic logistics processes to achieve the quantum leaps in performance being demanded of them. A variety of mechanisms are available to meet these new demands. They are as follows:

- **Reconfiguration**—Many companies are reducing the number of warehouses in their network. Most food producers, for example, used to have 11 to 14 warehouses to serve the country; today they have 6 to 8. If you have not assessed your warehouse network configuration recently, you may be missing an opportunity.

- **Restructuring**—In addition to reducing their warehouse networks, many companies have changed their distribution facilities' missions. Specifically, they have changed their inventory deployment strategies by centralizing slow movers and using cross-dock facilities to merge orders and to serve distant markets.

- **Third parties**—In many cases, shippers have reduced their logistics costs and asset bases and enhanced their flexibility by using third-party logistics service providers. These services must be evaluated on a case-by-case basis but may be a viable option for your company.

- **Carrier management**—Thoughtful shippers have changed how they procure transportation services, generally reducing the number of carriers with which they work and segmenting purchasing practices to utilize leverage where appropriate and to develop win–win partnerships where they make sense. The result has been better service and a lower transportation bill.

■ *Advanced information systems*—World-class shippers have continued to substitute information for assets and inventory and to use information to achieve operating efficiencies. Advanced information systems allow shippers to reduce costs *and* improve service. The necessary technology includes decision-support systems to improve planning, advanced applications software to enhance control, and real-time transaction data capture to improve both planning and control.

■ *Quality*—Finally, shippers are instilling quality in their logistics processes. This doesn't mean listening to spaced-out, impractical zealots pontificating about some ideal. Rather, in successful companies, it means awareness, understanding, and sensitivity to customer needs; attention to underlying business processes; and commitment to continuous improvement.

Logistics and transportation managers cannot escape the dilemma of having to deliver more for less. Customers are demanding it, and competitors are working to provide it. However, by moving beyond traditional thinking and by fine-tuning their current procedures, they can reach a new level of performance.

3.4 Differentiated Distribution

January 1987

An increasing number of companies are awakening to the concept of differentiated distribution. This philosophy recognizes the inherent differences in how customers want to be served and focuses a company's logistics systems accordingly.

Most companies sell to a highly differentiated market where the service needs and cost sensitivities of individual customers are different. Yet most companies don't recognize this fact and try to serve all customers as a homogeneous group; they offer common service levels, delivery times, and transportation means, as well as the same charges and prices. As a result, no individual customer group is served particularly well. Differentiated distribution, on the other hand, tailors your logistics system to meet the needs of specific customer groups.

In applying differentiated distribution, you begin with a customer service analysis detailing the service/cost requirements of various groups of customers (and potential customers) and products. Construct a matrix listing the product groups along one axis and the customer groups along the other axis. Then identify the service and cost needs of each customer/product group (that is, for each cell of the matrix).

The information on the customer/product matrix can help a company decide how to focus its logistics system. For example, a company might choose to focus on either cost-sensitive or service-sensitive market segments. Alternatively, a company might target a range of segments with different cost/service requirements. In this case, a dual distribution system, which utilizes different strategies or even different channels for serving each segment, can be effective.

One large supplier of veterinary supplies, for example, has successfully developed such a dual system by differentiating its service offerings along product lines. Veterinarians tend to stock lower inventories of vaccines. On the other hand, they maintain higher stocks of bandages, pest-control products, and other supplies. The veterinary supply firm, therefore, uses overnight air-parcel service for the service-sensitive vaccine product groups and cost-efficient ground transportation for other product families.

The importance of segmenting and targeting your markets is rooted in the concept of differentiated distribution. A logistics system gains extraordinary power when it is focused toward achieving a sharp goal such as high-quality service (within a loose cost constraint) or low cost (within a loose service constraint). The focus allows a company to align each element of its logistics system, producing a forceful competitive effect.

3.5   Value-Added Logistics

December 1988

Managed properly, logistics can be a powerful "value-adding" function. Most professionals working in the field readily understand that statement. They recognize that logistics excellence provides an avenue for gaining marketing advantage and for winning or retaining customers.

Nonetheless, too many companies today still view logistics only as a cost to be managed. They fail to recognize the inherent value of an excellent logistics system to the company and instead seek marketing advantage through product features, price, and promotion and advertising. These avenues are not always the best path to enduring competitive advantage for the following reasons:

- In many cases, ***product*** enhancements can be quickly matched by competitors. The pump toothpaste dispenser is a good example.
- ***Price*** reductions, without accompanying reductions in real costs, can be a losing tactic. As the passenger airlines have learned, competitors often quickly match price reductions, leaving the industry's competitive balance unchanged and everyone the worse off for it.
- ***Promotions and advertising*** rarely provide an enduring advantage. Without an accompanying price, product, or service advantage, promotions and advertising do little to contribute to long-term marketing success.

Value-added logistics, by contrast, can be an enduring avenue to competitive advantage. Structuring a logistics system that serves customers as they want to be served—providing on-time deliveries or ready information on order status, for example—usually provides substantial competitive benefits. Moreover, competitors cannot quickly duplicate these capabilities. It can take years to understand customers' needs and to design a logistics system with the appropriate warehouse network, information systems, organization, and operating effectiveness required to meet these needs effectively.

Some senior managers have recognized the power of value-added logistics; others still do not have a full appreciation of its potential contribution. Logistics practitioners can build awareness among top managers about the power of value-added logistics through enhanced customer feedback (for example, from formal customer surveys) and by bringing to their attention examples of successful companies that have utilized value-added logistics for marketing advantage. Building this awareness is essential for positioning logistics as a value-added function to be exploited rather than a cost function to be managed.

3.6   The Challenge of "Niche Distribution"

November 1988

Up through the 1970s, many U.S. companies designed their manufacturing and distribution systems to serve a mass market. In most cases, large volumes of uniform products were sold nationally through a defined set of wholesalers and retailers. Accordingly, the manufacturing and distribution systems were set up to minimize costs (assuming large production and shipment volumes), with little concern for flexibility.

Over the past decade, however, the rules of the game have changed. The mass market is rapidly becoming a memory of the past, and "local marketing" has become the wave of the future. Aided by point-of-sale scanning technology that captures sales data on a store level and faced with a proliferation of store formats with different logistics requirements, marketers have become increasingly sophisticated in segmenting customers and in targeting their marketing efforts.

This evolution to local markets presents challenges to logistics managers as they manage this new "niche distribution." Logistics systems now must contend with smaller shipment sizes to more outlets in a broader variety of channels. Furthermore, retailers are demanding higher service performance, often including elimination of stock-outs, more frequent deliveries, and increased value-added services, such as custom packaging, price tagging, or display building. Some also have asked their suppliers to provide flexible delivery systems with the capacity to deliver to the retail store, to the retailer's warehouse, or through a third party—frequently at no (or only moderate) increase over traditional costs.

Consequently, logistics managers have had to change the way they operate to meet the challenges of niche distribution. Most importantly, they have upgraded their command-and-control networks—the information systems that provide the capability to plan and track delivery requirements and inventory in the logistics system. Through the expanded use of electronic data interchange, bar coding of secondary containers, and new planning and inventory-control tools, logistics managers have shortened lead times, improved transaction accuracy, and deployed inventory more effectively. Furthermore, by developing channel systems that identify inventory levels and usage rates at the retail level, these companies have gained advance notice of sales trends and replenishment requirements.

The requirement for niche distribution is extending beyond traditional consumer goods channels. Logistics managers in all industries can learn valuable lessons from those segments that are successfully tackling these challenges.

3.7   Using Flexible Distribution as a Weapon

July 1988

Flexibility in distribution is becoming a key competitive requirement in many industries. Like low prices and superior customer service, flexible distribution (or the ability to serve customers in the way they want to be served) is emerging as a business necessity. Consider what has happened in the following industries:

- **Consumer packaged goods**—Mass merchandisers and grocery retailers are demanding expanded delivery capabilities from consumer packaged goods manufacturers. The demands are varied but may include the ability to drop-ship to individual stores rather than deliver to the customer's warehouse; arranging joint delivery and invoicing from multi-divisional manufacturers, which traditionally have operated independent distribution systems; providing value-added services such as inner packs, customized pallets, or electronic data interchange; and providing such extras as in-store merchandising services.

- **Automobiles**—Automotive assemblers have required suppliers to provide customized just-in-time delivery for many years now. Additionally, assemblers are now asking vendors to provide a broader range of value-added services such as sequenced delivery, bar coding, and assembled systems rather than just components (e.g., providing an air-conditioner system with all the belts, hoses, and connectors in place rather than just the base components).

- **Medical supplies**—Hospitals now require suppliers to deliver in more flexible ways and to provide such capabilities as daily delivery, use of a designated distributor, joint ordering and invoicing, and different delivery modes for different products.

These demands for more flexible and customer-responsive distribution provide new opportunities and challenges for manufacturers. Granted, creative approaches are needed to offer these extended capabilities at no (or modest) additional cost. Manufacturers that meet this challenge, however, will enjoy expanding business opportunities.

3.8   Time for a "Change-Based Strategy"?

July 1992

A new wave of thinking about business strategy now is evolving. Companies are beginning to realize that it is impossible to settle on a clear and permanent direction for their business because both the markets they serve and the environments they compete in are changing so rapidly. What's right for a company today may not be right for it tomorrow.

Leading companies, therefore, are developing "change-based strategies," that is, strategies that provide both the flexibility to serve diverse customer needs and the capability to respond to rapidly changing needs. This approach has profound implications for logistics and will make the discipline an increasingly critical business function through the 1990s. What follows is a brief history of business strategies adopted during the past few decades, followed by a description of the evolving "change-based strategy" approach and the role it will play in the future.

The broad acceptance of concepts known as the "growth/share matrix" and the "portfolio management" approach to business strategy heavily influenced most businesses in the 1970s. During those years, we used the labels *star, cow,* and *dog* to describe our businesses. Subsequently, frameworks and approaches like the "five forces model," "core competencies," "competing based on speed," and the "value chain" formed the core of strategic thinking in the late '70s and early '80s.

More recently, many companies have developed an appreciation of the value of "change-based strategies." The premise here is that most markets today are increasingly fragmented into smaller and smaller niches, each with distinctive requirements. These niches, moreover, have become increasingly dynamic, with the players in these markets (our customers) rapidly adjusting to the evolving needs of the customers they serve (the consumers). Therefore, our customers' needs not only have become more distinct, but also more variable.

For example, in the past, retailers could be accurately described by a class of trade—drugs, mass merchandise, and food. Today, these distinctions are meaningless, as retailers expand their lines, which themselves overlap to a great degree. Furthermore, these retailers have expanded the formats they offer considerably. As a result, manufacturers today face distinctive service requirements from convenience stores, warehouse-club stores, traditional supermarkets, "combo" food/drug stores, pipeline managers (such as Wal-Mart and Target) that want to build close operating linkages with their suppliers, innovators or marketers who want customized promotions, buying specialists who want to exploit forward-buying of promotional offerings, and so forth. As manufacturers try to serve these diverse accounts, they are overwhelmed by the different and complex service requirements.

Leading companies are responding to this dynamic environment by developing "change-based strategies." Their strategies normally include several key steps:

- **Understand the market**—Successful implementation of this strategy requires becoming adept at understanding evolving market directions and needs. Companies must work hard to remain close to their customers so that they understand their customers' strategies, directions, and requirements. Successful practitioners generally develop sophisticated ways to segment this market to understand the needs of their key customer groupings.

- **Translate to operating requirements**—Companies then must translate this understanding of the needs identified for each customer segment to specific operating requirements. Key capabilities for success are often flexible manufacturing, superior logistics, and leading-edge information systems.

- **Transform the organization and operation**—Finally, these companies will have to become expert in managing change. They must develop a management team that is effective in continually transforming the organization and operation to meet these rapidly changing requirements. Organizations that can instill effective change management as their core competence will be the winners in the 1990s.

Certainly, the logistics manager and logistics function will play a critical role in a "change-based" operation. Innovative logistics capabilities are needed to meet the cost, customer service, and flexibility demands placed by our customers in the 1990s.

Superior logistics involves several capabilities, including the ability to operate with minimal inventories, the flexibility to serve diverse customer delivery requirements (e.g., moving truckload lots to warehouse clubs and providing frequent deliveries of small orders to other accounts), and the ability to serve different lead time requirements. Other sought-after services include such value-added offerings as customized building of mixed pallets, ticketing, building assortments, specialized packaging, drop shipment, and direct store delivery.

Ironically, companies will be required to meet these new logistics requirements at the same or lower costs. As a result, leading-edge practices that promote operating efficiencies—such as cross-dock handling, sophisticated consolidation routines, pool distribution, core/non-core stocking policies, and computerized routing and scheduling—will become essential logistics capabilities in the 1990s.

STRUCTURING SUPPLY CHAIN CAPABILITIES

3.9	Time-Based Logistics

July 1989

The concept of a time-based competitive strategy is gaining increasing favor. Companies are finding that by reducing the lead times for all of their activities (manufacturing, moving products through the distribution channel, and order processing), they can obtain significant improvements in key competitive factors such as cost, quality, delivery times, and flexibility to respond to market changes.

Time-based strategies work because they eliminate uncertainty and waste from the system. In a traditional manufacturing operation, a product is actually worked on for only 2 to 5 percent of the manufacturing lead time! For example, if a product has a 20-day lead time, it likely is actually being worked on less than 24 hours. The remaining 19 days are spent in waiting, being moved, or being inspected or counted—all non-value-adding activities.

Many companies first applied time-based strategies to their manufacturing operations and now are extending these activities to other areas, including logistics. In doing so, they are finding that 50 to 60 percent of the total lead time is consumed by inbound and outbound logistics and order-processing activities. Thus, I believe that time-based logistics will receive increasing attention and that logistics managers should begin thinking about how to apply this concept to their operations.

To develop a time-based logistics system, managers need to concentrate on reducing lead times through a few key activities:

- ■ *Order processing*—Often, a careful analysis will uncover numerous opportunities to reduce the order cycle time significantly. For example, in many cases, orders are held by salespeople, communicated by inefficient means, and batched in lots for entry or for picking at the warehouse.
- ■ *Inventory movement*—Frequently, the velocity of inventory through the distribution channel can be improved. Multiple stocking points, multi-echelon warehousing networks with numerous intermediaries, and multiple handling of shipments slow inventory movement through the channel, lengthen forecasting horizons, and insulate companies from the ultimate demand point.

Becoming a time-based logistics competitor and focusing on reducing delivery lead times provide quantum benefits to a company. Although some of the steps needed to convert to a time-based system are counter to traditional thinking, doing so allows a corporation to reduce costs, improve quality, lower inventories, improve customer service, and increase operating flexibility. In fact, companies across a number of industries are already demonstrating the power of time-based logistics.

3.10 JIT Logistics

June 1988

As an increasing number of U.S. companies are learning, the just-in-time (JIT) inventory and management philosophy can pay big dividends. These dividends generally include lower costs—through the elimination of waste of all kinds—and improved quality (because companies can no longer hide problems). However, many companies are using JIT simply to transfer inventory to their suppliers, thus losing many of the real benefits of this system.

I find it useful to classify JIT programs into two categories, JIT production and JIT logistics. JIT production techniques represent the application of JIT principles within a manufacturing facility. These programs typically focus on the reduction of setup times for key operations, the reduction of lot sizes, and the enhancement of quality—all leading to lower work-in-process inventories.

JIT logistics programs, on the other hand, apply JIT principles to the management of raw materials inventories and inbound supplies. Too many manufacturers have not applied JIT logistics techniques properly, however, and have simply pushed inventory back onto their suppliers. If JIT logistics plans are to work, four "pillars" must be in place. They are:

- ***Stable production schedules*—A** manufacturer should provide suppliers with advance notice of the production schedule at least as far ahead as the supplier's processing lead time.

■ **Efficient communication**—Manufacturers and suppliers need to establish efficient communications and transaction-processing systems. Electronic data interchange linkages and paperless transaction systems are preferred because of their inherent speed and reliability.

■ **Coordinated transportation**—Distant suppliers can operate effective JIT logistics systems by working with other nearby suppliers to combine shipments. These "assembly and distribution" networks are becoming increasingly common and allow distant suppliers to make daily shipments economically.

■ **Quality control**—Source inspection at the supplier's plant or quality control systems that guarantee defect-free production are needed for JIT logistics programs to function well. Poor materials quality is the most common reason for the failure of JIT logistics systems.

Although they may sound self-evident, the four principles outlined above are critical to the integrated management of suppliers. Manufacturers that simply demand that suppliers provide daily deliveries and do not work actively with them to put these four "pillars" in place cannot realistically expect to reap the benefits of JIT logistics programs.

3.11 Channel Selection and Design— The Rediscovered Opportunity for Logistics

October 1994

The Logistics Strategy Pyramid (introduced in Section 2.14) highlights ten issues (organized on four levels) which must be carefully considered and integrated as a company develops a logistics strategy. When most companies develop a logistics strategy, they give little or no attention to the "channel design" component of the Strategy Pyramid.

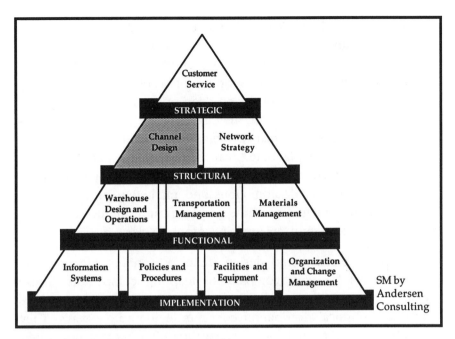

Figure 3.1 Logistics Strategy Pyramid

This component addresses critical issues such as:

- Should our company serve all or part of the market directly or should we use distributors?
- What are the costs, customer service impacts, and strategic implications of alternative channels?
- How can our company build channel linkages to gain cost and strategic advantage?

In recent years, a growing number of companies have reassessed the "channel design" component of the Logistics Strategy Pyramid with profound results:

- ***Consumer foods***—A major consumer products company which had considerable sales through small retail outlets (airports, hotel gift shops, etc.) had traditionally gone to market through specialty distributors. After a channel reassessment, the company switched considerable volume to a direct order service (800 number), with direct parcel delivery. This new "retail direct" strategy increased both profits and unit volume.

- **Medical products**—A supplier of a range of commodity medical products developed deep operating linkages with its major distributors. This strategy included continuous replenishment inventory management using "channel DRP (distribution resource planning)" capabilities, joint operations planning, "function-to-function meetings" with major accounts, and coordinated promotions and marketing programs. This "channel integration strategy" significantly reduced both logistics costs and channel inventory levels.

Channel selection and design strategies can provide powerful opportunities for your company to leapfrog competitors. These strategies provide the opportunity to link new marketing approaches *and* new physical distribution concepts in innovative ways. They are challenging to implement in that they often require new operating paradigms and new skills. However, they provide an opportunity to reinvent an industry or to profoundly change the economics of the industry in ways that provide your company with significant competitive advantage. Logistics managers can play an important role in the development and implementation of these strategies.

3.12 Third-Party Distribution

November 1987

An expanding array of third-party distribution services are providing new opportunities for logistics managers to reduce costs, improve service, and increase operating flexibility. These new services challenge traditional approaches but will nonetheless receive increasing acceptance in the coming years.

Third-party distribution services are not new. Manufacturers have long contracted out functions such as freight bill auditing, customs clearance, export packing and documentation, and public warehousing. Today, however, third-party services are no longer limited to these "nuisance" or "less significant" distribution activities. Many companies are now subcontracting important logistics functions as well.

Contracted traffic management services, for example, have replaced the traffic department in some companies. This practice has been common in Europe for some time, but it is just emerging as a viable option for the logistics manager in the United States.

In a few cases, shippers are even contracting for *total* logistics management services from third-party concerns. Reliance International of Lexington, Mass., for one, now markets its expertise in areas that include logistics network design, distribution planning, carrier negotiation, and ongoing traffic management and operational support. Competing with companies like Reliance International, moreover, are several carriers that have recently gotten into the act. Today, Leaseway and Skyway, among others, have broadened the range of logistics services they offer shippers.

From a shipper's point of view, the benefits of these services are potentially significant. They include lower administrative costs, increased flexibility, lower freight and warehouse rates through greater purchasing leverage and better network design, and improved management control.

Keep in mind that these services are not risk-free and that shippers will need to monitor the work of third-party operations to assure that their standards of quality are maintained. Nonetheless, third-party distribution services provide an interesting new option for logistics managers to consider.

3.13   The Perils of Partnerships

February 1992

The concept of supplier or customer partnerships has received enormous attention in recent years. Known variously as Quick

Response, Channel Integration, Inter-Corporate Logistics, and Supplier (or Customer) Connectivity programs, these partnerships are designed to reduce the significant inefficiencies built into the supply chain when suppliers and customers operate independently. If their operations can be coordinated and managed in an integrated way, the theory goes, both parties can achieve significant efficiencies.

The documented success stories of suppliers and customers that have benefited from channel partnerships are growing and impressive. Yet despite the hoopla, many companies are discovering that partnerships are not for everyone. Or to be more precise, companies benefit in wildly disproportionate degrees from partnerships. Certainly, each company must tailor a partnering strategy and approach that *is* right for it. If it doesn't, it may end up sorely disappointed with the results.

In theory, it's hard to argue with the concept of partnerships. The idea of eliminating the waste inherent in the supply chain is appealing. In practice, however, several things work against effective partnership relationships. To avoid these traps, it is important to recognize the following:

- Benefits fall disproportionately (or even fully) to the "supplied" (the customer). The supplier sees its costs (e.g., handling, transportation, and/or inventory) increase. Many retailers argue, "Let's work to build efficiency in the logistics channel and let the benefits fall where they may. In the end, you (the supplier) will benefit as well." Nice try, but no one's buying it!

- The investment costs (information systems, operations changes, and especially management time) typically exceed expectations and preempt or limit the partnership relationships a company is able to establish.

- The "propensity to partner" does not exist. Some companies have a culture and attitude that are not conducive to partnering. Don't waste your valuable time trying to work with an unwilling partner.

Having worked closely with numerous companies in formulating partnership strategies, I've concluded that each organization should develop an approach that is best for its particular circumstances. Still, it is possible to draw a few generalizations about the following groups:

- *Large retailers*—Of all the groups described here, large retailers tend to be the most interested in having broad partnership relationships. Not only are they in a position to realize tremendous inventory savings and operating cost reductions, but they also are becoming increasingly capable of forcing a "partnership" on their terms.
- *Large suppliers*—Big manufacturers and distributors are in a position to realize substantial benefits from partnerships—particularly if they are branded, non-seasonal, category-dominant suppliers of logistics-intensive products.
- *Mid-sized suppliers*—These companies must craft their partnership strategies carefully. They must focus on the best opportunities, select partners carefully, choose tactics appropriate to each individual situation, and preempt/lead partnership efforts (that is, avoid being forced to follow their customers' prescribed programs).

I do believe that appropriately designed channel partnerships can add tremendous efficiency to the distribution channel and reward all participants with lower costs, better service, and high sales. But there's a caveat here: If not carefully designed, partnerships can end up like a bad marriage, with each party losing something from the experience.

3.14 The Emergence of Manufacturer Alliances

May 1992

The development of partnerships has received considerable attention in recent years. Companies across a broad range of industries have achieved significant benefits by changing how they work with their suppliers, carriers, and customers. In past columns, I have described how companies gained dramatic strategic and operating advantages by building partnerships with these external parties, a process I refer to as channel integration.

More recently, manufacturer alliances have begun receiving attention as well. These alliances involve non-competing manufacturers collaborating to gain cost and customer service advantages. They are fundamentally different from channel integration partnerships, which involve the "vertical" integration of activities across the supply chain.

Manufacturer alliances are attracting interest for several reasons. With the shift in power to the end of the channel (e.g., to the retailer or user) in most industries, manufacturers are being squeezed and required to deliver more for less—that is, provide increased customer service at lower costs. Even though most manufacturers have limited latitude to respond, they still must meet the retailers' demands (and experience higher costs and lower margins) or see their business volume erode.

Manufacturer alliances provide an additional way for manufacturers to deliver more for less. Strategically selected alliances, moreover, can increase the manufacturers' leverage with customers.

Manufacturers have several options to consider when selecting alliance partners. Figure 3.2 outlines a framework for manufacturers to use in identifying potential partners.

Despite the myriad possibilities, there are strikingly few examples of successful and enduring manufacturer alliances. The 3M/Abbot alliance in the healthcare products industry is probably the best example of an alliance that has survived the test of time.

Alliances are torpedoed or never get off the ground because parties are unable to navigate through several obstacles, including:

- The business practices of the companies are incompatible or unable to be reconciled.
- Companies are unable to compromise on the degree of control they want to exert over the alliance operations.

	Product Physical Characteristics	Demand Characteristics	Alliance Characters		
			Logistics Complement	Channel Expansion	Category Complement
Points of Leverage	• Complementary "value density" of products • Opportunity to optimize weight and cube	• Consolidate LTL shipments • Build more TL shipments or higher weight breaks	• Complementary seasonality • Complementary promotional cycles	• Leverage channel coverage	• Leverage buying, merchandising, and delivery of products sold in same category
Benefits	• Reduced shipping costs • More frequent delivery	• Reduced shipping costs • Faster transit times • More frequent delivery	• Smoothed peaks • Lower operating costs	• Broader distribution • Higher sales volume	• Better coverage • Lower cost • Increased sell-through
Illustrative Example from Consumer Packaged Goods	• Cereal and soap	• Baby food and condiments • Cigarettes and candy	• Antifreeze and suncare products	• Detergents and bandages	• Paper towels and diapers

Figure 3.2 Framework for Identifying Manufacturing Alliance Partners

- The companies do not have the information systems or link-ages to make the alliance work.

Nonetheless, over the next five years, we will see more com-panies overcome these obstacles and build successful alliances. The operating and strategic benefits of these relationships can be compelling, and they will be used to help manufacturers maintain their margins in the face of increasingly powerful and demanding customers.

3.15 The Manufacturing Connection

December 1995

In recent years, logistics professionals have turned most of their attention to the logistics/customer interface. Among the important questions they have confronted are:

- What are our customers' needs and how can logistics better support them?
- How can we build channel linkages with customers and distributors?
- How can we create better links with marketing and sales to enhance forecast accuracy?
- What is our profitability for each product and for each major customer or customer segment?

These are all important issues. While addressing them, however, some logistics managers lose sight of the "logistics–manufacturing connection." That is, they fail to realize how tightly integrating logistics and manufacturing can drive costs down and improve return-on-asset performance. Linking these areas also can signifi-cantly enhance flexibility and customer service performance while addressing the customer-related issues noted above.

Specifically, I see three areas where alignment of logistics and manufacturing strategies can have a big impact:

■ *Flexible manufacturing*—In spite of the tremendous advances that have been made in setup reduction, cellular manufacturing, and implementation of world-class manufacturing strategies, a very large number of companies still run their manufacturing operations with long production runs that are designed to minimize the number of setups and changeovers and to reduce manufacturing costs as much as possible.

Unfortunately, too many manufacturing managers today are measured primarily on manufacturing cost variances. As a result, they are selling their companies' souls by mandating long production runs in order to achieve a narrow and inappropriate goal.

Every thinking individual who has seen flexible manufacturing at work comes away a convert. Flexible manufacturing enhances forecast accuracy, significantly improves customer service, chops unproductive inventories (both finished goods and work in process) by 50 to 90 percent, enhances cross-functional coordination, reduces total costs, and aligns production with marketplace realities.

■ *Warehouse conversion*—Increasingly, companies are finding that not all manufacturing has to occur in the plant. Many distribution centers today are performing operations such as assembly, final configuration, packaging, construction of promotional displays, and merging components. These activities leverage the concept of "postponement" and can yield significant cost, customer service, and inventory advantages.

■ *Structural alignment*—Opportunities exist for most companies to enhance performance through structural alignment of manufacturing and logistics. The impact of this tactic is magnified as we move toward global operations. Structural alignment requires us to consider such questions as the following:

☐ Does reduced-scale, regional, or local manufacturing make sense?

☐ Which products should be produced at which facilities? Should all facilities make all products?

- ☐ How many distribution centers should we utilize and where should they be located?
- ☐ Can we utilize cross-dock or merge capabilities?
- ☐ Should our distribution centers be plant- or market-based?
- ☐ Can we do more plant-direct shipments, bypassing the warehouse altogether?

The logistics profession has made tremendous advances over the past decade, including developing a heightened awareness of customer needs and a commitment to customer service excellence. We have enhanced channel and customer linkages to great benefit. However, we must not lose sight of our roots. By leveraging a tight integration of logistics with manufacturing, we can assure optimal supply chain performance.

3.16   Time for Quantum Change

April 1989

Which type of change is easier for operating managers to cope with—sudden change or gradual, evolutionary change? The answer may surprise you.

In many ways, sudden changes are easier to manage than changes that develop slowly over an extended period of time. The need for a rapid and profound change is almost always apparent, and the clearer the need, the easier it is to mobilize an organization to act.

Gradual, evolutionary change, on the other hand, can be quite insidious for operating managers. As operating needs, product lines, technological options, and the markets served gradually change, managers must alter their operations accordingly. However, the need for a change—and a determination of what direction the change should take—is not always clearly apparent to the organization. And even when companies recognize the need for change, there is a natural tendency to resist.

The following examples illustrate what I call "the evolutionary trap," which can often unsettle the operations of even the best-run organizations.

- A manufacturer of oil-drilling equipment sold standard items from inventory. As equipment needs gradually became more specialized, the engineering content required for each order increased and the manufacturing mission became considerably more complex. Because the changes occurred gradually, the company was slow to recognize that producing "customized" goods required different capabilities—different order-entry systems, engineering capabilities, inventory-control methods, and manufacturing systems—from those needed for selling "standardized" products.

- As the products offered by many computer companies changed from large mainframe systems in the 1970s to include more minicomputer and microcomputer systems in the 1980s, many companies were slow to modify their logistics systems. Forecasting methods, order-entry and -processing capabilities, and inventory management requirements are different for selling a single million-dollar mainframe than for selling numerous $2,000 microcomputers.

These "evolutionary traps" are difficult to avoid, but successful companies generally recognize the need for change before the symptoms of longer order cycles, missed shipment dates, higher costs, and customer dissatisfaction appear. These companies have learned to step back from day-to-day operations periodically, assess their businesses, identify strategic requirements for operations and logistics, and install operating systems needed to support the new business requirements.

4 CUSTOMER SERVICE

L abeling the 1990s the "Decade of Customer Service" is not mere rhetoric. Companies are feeling increasing pressure from customers to deliver higher levels of customer service. Recognizing this trend, the selections in this chapter emphasize the need to base logistics and supply chain programs on customer needs, not on internal factors. This chapter outlines ways to understand the customer and to incorporate that understanding into supply chain and logistics management.

The ten columns in this chapter fall into two clusters:

- The first four selections focus on understanding the customer. They offer suggestions on how to think about customer service (no, reliability is not enough), how to obtain customer feedback on service, and how to segment customers to help enhance service. "Using Customer Segmentation for Logistics Advantage" (4.4—April 1994) provides guidance in how to proactively move beyond a "one size fits all" approach to customer service.
- The other six columns (4.5 to 4.10) concentrate on translating understanding of the customer into enhanced capabilities. Topics range from setting a customer-driven logistics strategy to ensuring the "perfect order" to linking logistics with the order management process. "Matching Logistics to Customer Needs" (4.6—September 1990) is the only column in this book not published by *Logistics Management*. It was a guest column which appeared in *Distribution Magazine*. "Creating the Perfect Order" (4.7—February 1993) generated considerable interest and en-

couraged many companies to try to develop a single, more comprehensive measure of customer service.

UNDERSTANDING THE CUSTOMER

4.1   Pyramid of Customer Service

August 1991

The 1990s have been called the "Decade of Customer Service." Companies in virtually all industry sectors are placing a premium on quality, including quality customer service. "Serving our customers as they want to be served" and "making our company easy to do business with" are competitive objectives for the 1990s.

This focus on quality and on the customer has important implications for both shippers and carriers. To be successful, both must develop the capability to serve their customers effectively. At the same time, however, the meaning of effective customer service is changing, and companies must meet an increasingly higher standard. To help formulate a customer service strategy in this fluid environment, I have found the Customer Service Pyramid to be an effective framework.

The Customer Service Pyramid segments customer service capabilities and programs into three tiers, as shown in Figure 4.1. At the base of the Customer Service Pyramid is *reliability*. Achieving a reputation for reliability means performing the basics well—that is, maintaining short order cycle times, delivering on time consistently, processing paperwork accurately, and making damage-free shipments. In the past, effective performance on the *reliability* tier defined excellence in customer service. Today, this is no longer true. If a company only does the basics well, it will likely have a difficult time just maintaining market share. Companies must develop programs on the other customer service tiers.

The second tier is *resilience* or the ability to respond to failures of the customer service systems. To achieve resiliency, a company

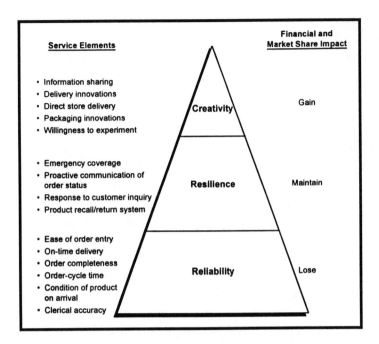

Figure 4.1 Customer Service Pyramid

must offer an effective rush-order shipment capability, provide proactive communication of changes in order status, maintain effective shipment-tracking capabilities, and immediately respond comprehensively to customers' inquiries.

The top tier of the Customer Service Pyramid is referred to as *creativity* or *innovation*. Demonstrating creativity means developing value-added programs for your customers—such as drop shipments to store locations, customized pallet building, innovations that reduce handling costs, and services tailored to an individual customer's needs. It involves a willingness to experiment and an operating flexibility to create programs that add value.

Companies must develop programs in all three tiers. Clearly, they must have the base or *reliability* in place. But to build a differential position in the marketplace that increases market share, they also must develop capabilities and programs at the *resilience* and *creativity* levels.

4.2 | The Customer Service Survey

May 1987

A focused logistics and customer service strategy that is supported by all parts of the organization is still only a dream in many companies. That's because these companies have not developed and articulated clear, consistent customer service goals. This often leaves the logistics function unfocused and vulnerable to unclear or changing demands.

The first step a logistics manager can take to achieve a focused strategy is to develop a well-conceived customer service survey. Most companies either don't conduct customer service surveys at all or, if they do, simply include one or two vague questions about customer service as part of broader marketing surveys. Neither approach is sufficient. A complete customer service survey is needed; it should accomplish the following:

- Provide a quantitative understanding of the customer requirements for each element of customer service (delivery speed, delivery reliability, order completeness, product availability, etc.)
- Measure the relative importance of each element of customer service
- Assess the performance of your company and of your major competitors for each element of customer service
- Provide an understanding of the relative significance of customer service issues in your customers' overall buying decisions (i.e., vs. price, product features, product quality, and so forth)

A task force involving all corporate functions should work with the logistics manager in determining the survey approach. The task force must consider such important issues as the survey vehicle (telephone, mail, in-depth customer interviews, or a combination of these), the survey target (who in the customer organization is the key decision-maker or an important influence), the definition of customer and product groups, the survey questionnaire design, and how the survey responses will be analyzed.

The logistics manager, as well as the corporation as a whole, can benefit greatly from a properly conceived and executed customer service survey. Such a survey identifies the strategic focus of the logistics system; provides specific requirements and design criteria for that system; facilitates communication and harmony among all corporate functions, particularly between logistics and sales/marketing; provides valuable competitive information; and opens valuable communication links with customers. Perhaps most importantly, the survey can provide a clear direction to the logistics function, allowing it to focus its full energies on creating and managing a system that most efficiently achieves the consensus customer service goals.

4.3 The "Transaction Survey"

November 1991

Effective customer service has become a competitive requirement for the 1990s. Both shippers and carriers now realize that to attract and retain business, they must meet increasingly stringent customer service requirements. And they are discovering that clients are willing to pay a premium for superior service. More than one carrier, for example, has been able to ameliorate the intense price squeeze by positioning itself as a superior service provider.

As part of their efforts to achieve customer service excellence, carriers and shippers have initiated a variety of programs. These include customer surveys and focus groups designed to help them understand customers' service needs and expectations; quality programs to sensitize employees to the new reality and to assure compliance with customer expectations; and investments in information systems, training, and enhanced delivery capabilities to meet the new service standards.

In addition, several shippers and carriers have instituted the

Transaction Survey to supplement these efforts. The Transaction Survey is simple and inexpensive to execute, yet it is a very effective way to stay in touch with customers and to monitor their needs and your company's compliance with their expectations.

Here's how it works. You keep track of the many "transactions" your customers have with your company as they order, receive shipments, inquire about order status, and call with questions or complaints. At the end of the day (or the very next day), you select a sample of 2 to 5 percent of each of these transaction types and call your customers to ask a few simple questions along the following lines:

> Your company was involved in a transaction with our company today. You placed an order for 30 boxes of 6 different components (or we delivered XXX to you). How was that transaction in general?
>
> - Did we meet all your needs?
> - Was the order handled quickly and effectively?
> - Was the customer service representative (or driver) courteous and responsive?
> - How could we have met your needs better?
> - Are competitors more effective in filling your orders?

You then tabulate the results and provide your management team with a current, accurate picture of your company's customer service performance. This method of customer monitoring works because it deals with a recent event. Unlike the annual customer service survey, in which customers are queried about transactions that may have occurred months ago, responders to the Transaction Survey are asked only about specific events that happened less than 24 hours ago. They are not dealing with abstractions or impressions. Therefore, the feedback is more accurate and reliable.

Furthermore, results are processed essentially in real time so that you can deal with problems promptly. In combination with a comprehensive annual customer service survey, the Transaction Survey provides the foundation for excellence in customer service.

4.4 Using Customer Segmentation for Logistics Advantage

April 1994

Logistics managers can provide considerable value to their companies by understanding their customers' delivery requirements. A useful and very powerful tool for understanding these requirements is *account segmentation.*

Traditionally, most companies have either viewed their customers as a monolithic entity or they have divided their customers into groupings based on such distinctions as "class of trade." In the consumer packaged goods industry, for example, many manufacturers group customers according to the type of business they are in—grocery, drug, or mass merchandising. As differences between the various classes of trade have blurred, these distinctions have become less meaningful for manufacturers to use as the basis for understanding customer requirements and for organizing their companies to serve these customers.

Now, companies are finding other, more useful approaches to grouping their customers. One of these is account segmentation, which organizes customers into common groupings based on key attributes. Through creative account segmentation, companies can find new ways to think about their customers. A company can use account segmentation to identify market segments that it is well positioned to serve and then organize its product and service offerings to serve them in a distinctively superior way. As part of this process, the company must tailor its logistics offerings to meet the needs of each of the market segments it serves.

One consumer products manufacturer, for example, segmented its customers based on their operating sophistication and merchandising philosophies. Customers' operating sophistication was determined by the degree of development of their information systems and operating methods. Their merchandising philosophies included such sales approaches as "every-day low prices" and "deal-oriented" (a heavy emphasis on discounting and promotions).

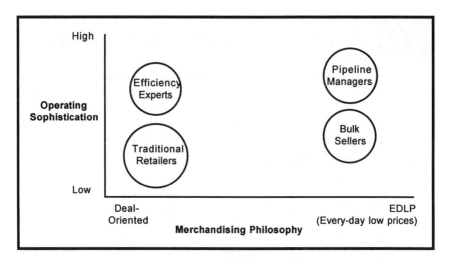

Figure 4.2   Segmentation of Consumer Packaged Goods Retailers

Analyzing customers according to these characteristics helps to identify segments—that is, groups of customers with common needs—that are distinct from one another. Figure 4.2 is an example of a segmentation of consumer packaged goods retailers.

The customer service requirements and demand for logistics and marketing services differ for each of the segments in Figure 4.2. In essence, the retailers in these segments have different strategies for success and demand different things from their suppliers. For example, a "pipeline manager" (a Wal-Mart, for example) has different in-stock availability, lead time, and EDI linkage requirements for its suppliers than does the traditional retailer. Similarly, a "bulk seller" (a Costco-type buying club) requires different shipment configurations and delivery requirements from the other segments. "Efficiency experts" (such as Food Lion) focus on purchasing and delivery efficiencies and, therefore, have yet another set of logistics service requirements.

Logistics managers and their companies will be most successful if they recognize these important operational differences and service requirements among their various customer segments and orient their capabilities toward the needs of each distinctive segment.

BUILDING A CUSTOMER-FOCUSED SUPPLY CHAIN

| 4.5 | Developing a Customer Service Strategy |

April 1987

Many companies today still operate without clear statements of mission and without specific performance targets for their logistics function. The absence of these goals and targets complicates the job, causes unnecessary conflict and confusion within the company, and often leads to a logistics system that is not in harmony with corporate strategic objectives. Development of clear goals and specific performance targets will benefit both the logistics function and the company as a whole.

Without goals and performance targets that are agreed to and understood by all parts of the company, the logistics function is vulnerable to random demands from other parts of the organization. One month the emphasis may be on lowering inventories, the next month it may be on improving order-fill-rate performance, then there may be pressure to reduce freight costs, and subsequently the demand may be to shorten delivery times.

Even more disturbing than these changing demands is the insistence that all of the activities be done well. Although not often using these exact words, some companies ask their logistics functions to "provide the best service in the industry at a cost lower than all competitors."

As nice as that sounds, it is not possible to achieve. There are unavoidable trade-offs between cost and service. Too often these realities are ignored and the logistics function becomes the "fall guy" for poorly focused business and marketing strategies. Demands are capriciously made to improve "customer service" without knowing which elements of service to improve and how much they need to be improved.

Some of these confusions, conflicts, and missed opportunities can be avoided through explicit goal and performance target setting. Prepare a clear, written statement of the logistics mission. Make certain that this statement is consistent with key corporate strategic thrusts. Then obtain consensus on it from all parts of the organization.

Subsequently, translate the statement of mission into specific, quantitative cost objectives and customer service goals for each element of customer service. Then, obtain organizational consensus on the performance targets, use these targets as the guidelines for designing and fine-tuning the logistics system, and track actual performance against the targets.

Developing and using these specific targets will focus the logistics function on the critical tasks, reduce organizational conflict, and build a team united in achieving customer service goals.

4.6 Matching Logistics to Customer Needs

September 1990

Many companies describe themselves as customer-driven, focusing on serving their customers the way their customers want to be served. To be truly customer-oriented, companies need to build leading-edge logistics systems. To do that, these five areas must be addressed:

1. ***Understanding customer service needs***—This is the starting point for differentiating your logistics capability.
   - Which customer service elements are important to customers? Are different elements important to different customers?
   - How are we performing on each element of service—from our customers' perspective? What performance levels are acceptable? How do we compare to key competitors?
   - What value-added capabilities can give us a distinctive edge?

2. ***Structure and operating policies***—Many companies can gain significant cost savings and customer service improvements from a redesign of their logistics network. Use a logistics network planning model to address basic questions such as:

   ■ How many distribution centers should we have? Where should they be? How big should they be?

   ■ What are the cost and customer service implications of alternative logistics network designs? What logistics network configuration makes the best strategic sense?

   In addition, a comprehensive assessment of your logistics policies and procedures can uncover opportunities to improve efficiency.

3. ***Organization***—Integrated logistics requires careful design of three elements:

   ■ *Organization structure*—There isn't a single structure that is best for all companies. The assessment of responsibility for logistics functions must be determined by considering key competitive factors, customer needs, and the culture and philosophy of each company.

   ■ *Roles and responsibilities*—In a stable environment, let each function manage independently, with the goal of achieving the greatest functional efficiency.

     In an uncertain or dynamic environment, develop an organization which has close working relationships among functions and operates in an integrated way.

   ■ *Performance measures*—Align performance measures with desired results. Don't ask a purchasing manager to lower the total cost of a purchased item if he or she is evaluated solely on purchase price. A transportation manager will not set up the most efficient logistics system if bonuses are based on freight bill reduction targets alone.

4. ***Information systems***—Effective information systems are critical to successful, customer-driven distribution. Three aspects

of these systems are important. First, timely and accurate information is essential. Second, integrated applications software with full functionality is a key part of a successful logistics information system. Third, advanced decision support is a valuable tool for the logistics system manager. Capabilities in this area include logistics network planning models which allow a "what if" simulation of the cost and customer service impacts of alternative logistics network structures and policies.

5. *Channel integration*—For many companies, quantum improvements in logistics performance are no longer available by fine-tuning their own systems, but by integrating their logistics system with those of their suppliers and customers. It makes no sense for a manufacturer and distributor to hold inventory in warehouses next door to each other. If they can manage the inventory jointly, they can improve the performance of the entire logistics channel.

4.7   Creating the Perfect Order

February 1993

A new, simple, and comprehensive measure of customer service and logistics effectiveness is being employed by a small but rapidly growing group of companies. It is called the "Perfect Order."

The Perfect Order is designed to measure the effectiveness of a process, not the effectiveness of a particular function. In simple terms, it measures the percentage of orders that proceed through every step of the order management process without fault, exception handling, or intervention.

Each step in the order management process must go smoothly for the order to be considered perfect. These steps include order entry, credit clearance, inventory availability, accurate picking, on-time delivery, correct invoicing, and payment without deductions.

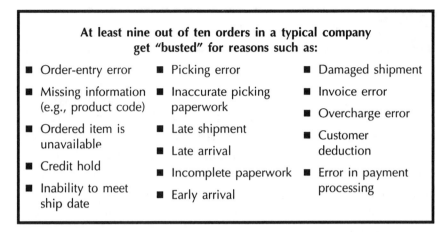

At least nine out of ten orders in a typical company get "busted" for reasons such as:

- Order-entry error
- Missing information (e.g., product code)
- Ordered item is unavailable
- Credit hold
- Inability to meet ship date

- Picking error
- Inaccurate picking paperwork
- Late shipment
- Late arrival
- Incomplete paperwork
- Early arrival

- Damaged shipment
- Invoice error
- Overcharge error
- Customer deduction
- Error in payment processing

Table 4.1 **"Perfect Order Busters" (selected examples)**

If anything goes wrong (or requires manual intervention, exception processing, or expediting), the order is not considered perfect. (See Table 4.1 for a list of problems that can sink an otherwise perfect order.)

Companies are beginning to measure the percentage of their orders that meet the criteria of the Perfect Order. Surprisingly, they are finding on average that less than 10 percent of their orders are perfect. As a result, these companies are re-engineering their order management and logistics processes to eliminate unnecessary steps, align functional objectives, and speed up order cycles. They are using the Perfect Order to measure their success.

One major consumer packaged goods company has achieved remarkable success in this regard. It has increased its Perfect Order percentage to between 55 and 60 percent, and its goal is to reach 90 percent within two years. Because its competitors' effectiveness is just one-sixth of this rate, it has a growing competitive advantage.

Measuring the effectiveness of processes will become commonplace in companies in the 1990s. The Perfect Order is a sound measure of order management and logistics effectiveness, and it provides a lever for companies to improve customer service and reduce costs at the same time.

4.8	Building a Trade Franchise

July 1991

The concept of a "brand franchise" is well understood. Typically, consumer products (and industrial products) companies strive to create this type of compelling demand for their products through effective product design, positioning, and advertising. Notable successes include such brand names as Tide, Crest, Ivory, Listerine, Sony, Caterpillar, Maytag, and Kitchen-Aid. To a one, these brands connote value and images of quality, for which consumers willingly pay a premium.

Though less well understood, the concept of the "trade franchise" also is beginning to provide significant benefits for the selected companies that have concentrated on its pursuit. In a phrase, building a trade franchise means making your company easy to do business with. It involves all the activities of your company, but focuses prominently on logistics. For example, to build a trade franchise, your company must:

- Make it a priority to provide superior customer service so that your company's products are rarely—if ever—backordered.
- Maintain the flexibility to serve your customers as they want to be served. For example, your company might accommodate a customer's preference for delivery to the warehouse, the retail store's back door, or even the retail shelf. Similarly, a multi-divisional company might develop the capability to accommodate its customers' preference for a single order and single shipment for all divisions' products.
- Actively explore ways to reduce your customers' costs through the use of bar coding, appropriately sized inner packs, or by building mixed pallets to reduce your customers' handling requirements.
- Tailor promotions to individual customer needs.
- Provide easy and efficient ordering, order inquiry, and problem resolution capabilities.

In short, a trade franchise involves working with your customers in a partnership fashion in order to maximize sell-through and to minimize joint costs.

Though the trade-franchise concept already has begun demonstrating its merits, its value will increase significantly in the 1990s for the following reasons:

■ Retailers' power will increase as consolidation in the industry continues and as retailers begin to exploit their control of the key source of market information (point-of-sale scanning data). Retailers will exercise this power through more direction of promotional and marketing programs (targeted at the individual store level) and through more demands on suppliers.

■ It will become more costly to build a brand franchise because of the dilution of the mass media (e.g., 50 to 500 television stations on the dial today vs. 3 to 8 ten years ago) and because of the high costs of advertising.

Suppliers that build trade franchises that make them easy to do business with will gain a preferential position. Specific advantages will include preferred shelf positioning and space, easier retailer acceptance of new products, active promotional program compliance and support, and better product availability. These advantages translate into higher sales volumes and greater profits.

Effective logistics performance is central to the development of a trade franchise. The logistics manager's job, therefore, will increase in importance as he or she takes a leadership role in developing a trade franchise.

4.9	Assessing the Order Management Process

April 1995

Although many companies have not always linked the order management process with logistics, I believe that they must explicitly consider this process as part of the development of a logistics strategy. While it may be organizationally separate from logistics, the order management process must be assessed as one of the variables affecting supply chain performance. For example, it conceivably could be easier and less costly to reduce the order-processing cycle time by one or two days as a way to shorten delivery times than to add distribution centers closer to key markets.

The order management process involves all the activities from initiation of customer orders through order processing, order confirmation, order picking and shipping, delivery, billing, collection, and reconciliation of any invoice deductions. Most companies that analyze their order cycle times are surprised to find that the average order cycle time and the range of order cycle times will vary considerably. The analysis in Figure 4.3 (taken from a sample of 200 customer orders) is typical of many companies and reflects statistics from a large consumer packaged goods company.

Surprisingly, the company found that order cycle times averaged more than 12 days. Although it was able to expedite orders to serve some customers in as little as 5 days, the 95th percentile of order-processing lead times was over 32 days, meaning that almost half of the orders were taking from 12 to 32 days to process and deliver to customers. Furthermore, many companies find that the administrative burden required to fully process and close an order can be considerable. For example, one $500 million consumer products company employed more than 150 people for these activities.

Completely redesigning or "re-engineering" the order management functions can shorten order cycle times, improve the consistency of order cycle times, reduce administrative costs, and enhance customer satisfaction at the same time. Here are some tips on how to proceed:

- Use the broad definition of order management suggested above. Think beyond just order entry or order processing.
- Be bold. Set stretch targets of quantum improvements, not incremental change.

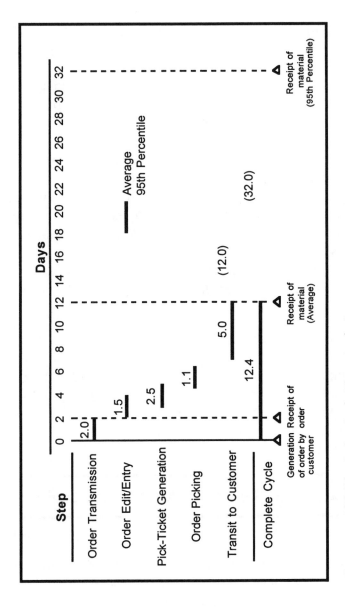

Figure 4.3   Sample Order Cycle Time Analysis

- Consider organizing in "cells," where a small group of people is responsible for all the order management activities, from order entry through deduction processing. For example, one company defined these cells by type of order—fax orders, EDI orders, and continuous replenishment orders—and organized teams for each type of order.

- Consider the big picture. For example, determine whether complex promotion and pricing practices could be contributing to errors and the length of the order cycle.

- Relentlessly seek speed. Eliminate dead time, non-value-added activities, and waiting time, where an order sits and nothing happens. Design the process so that the orders flow.

- Leverage technology where appropriate. Use technology to speed the process, to provide the integration information needed for effective order management, and to leverage human skills.

- Utilize measures like the "perfect order," which assesses the percentage of orders that are error-free and do not require manual intervention.

Companies have pared order cycle times from weeks to days or hours—while reducing administrative costs by 10 to 30 percent—by re-engineering the order management cycle. Furthermore, these companies have integrated order management and logistics into an aligned supply chain strategy to meet expanding customer requirements, while lowering costs at the same time.

4.10 Customer-Driven vs. Asset-Driven Logistics

April 1995

Almost every company I come into contact with today likes to describe itself as being "customer-driven." These corporations believe they are in touch with their customers' needs and that they have oriented their operations to be responsive to those needs.

Although these companies are sincere in their perception of themselves, few of them are truly customer-driven, in my opinion. Rather, most are asset-driven, having focused their executives' time and attention, their measures and rewards, and their management control systems on internal rather than customer-based parameters.

Take the test below to find out if your company really is customer-driven.

Customer-Driven Quiz

	Asset-Driven	Customer-Driven
	1 2 3	4 5
Amount of time senior management and functional executives spend meeting with customers	Once/twice per year	Greater than 20% of time
Conduct formal customer service and customer satisfaction research	Rarely	Yearly
Conduct formal multi-functional meetings with customers	Never	Multiple times per year
Formally measure customer service performance	No	Formal tracking on weekly basis
Use customer satisfaction as key measure of executive performance	No	Yes
Track equipment utilization as key measure of performance	Yes	Not critical factor
Focus on cycle time reduction as key management parameter	Not at all	Critical dimension
Management meetings allocate time equally to internal processes *and* external issues such as customer needs and competitor actions	Internal focus	Balanced focus
Conduct joint problem-solving meetings with customers	Never	Routinely

If you score at least 35 points on the hypothetical scale, you have earned the right to call yourself customer-driven. Scoring fewer than 20 points implies that you could significantly enhance your relationship with your customers.

Logistics managers can play an important role in assuring that their function—indeed, the entire company—remains customer-driven. Increasingly, companies are differentiating themselves from their competitors on the basis of logistics excellence. Achieving that distinction requires a thorough understanding of customers' logistics requirements and demands that logistics managers stay close to customers to keep abreast of their changing needs. It is essential that they manage their activities so they are aligned with those customer needs.

5 FUNCTIONAL EXCELLENCE

This chapter argues that the key to functional excellence is recognizing all the activities involved in managing the supply chain and the importance of excellent execution of each individual activity as well as the totality. Thus, the whole becomes greater than the sum of its parts.

The first column in this chapter shows how a number of activities link together to form a logistics process. Then, the other 17 columns examine how to build superior capabilities in each of the four groups of activities that constitute this process:

- Seven selections (5.2 to 5.8) focus on forecasting and inventory control as tools for planning and managing resources to support superior customer service. The evolution from "Time to Consider DRP" (5.5—September 1989) to "The Emergence of Channel DRP" (5.6—June 1992) and "Stages of Continuous Replenishment" (5.8—August 1993) reflects both the simplification of distribution center networks over the 1980s and the emergence of extended supply chain relationships in the 1990s.
- The next six columns (5.9 to 5.14) examine new approaches to achieving excellence in transportation and warehousing, including the difference that quality improvement programs make and new operating principles that are transforming distribution centers. "Should You Switch to Cross-Dock?" (5.14—September 1990) generated considerable interest and reflected the emerging trend to eliminate stocking locations, enabled by new transportation

services and enhanced information system capabilities and planning tools.

■ The next two selections (5.15 to 5.16) integrate purchasing into the supply chain and discuss new approaches to improve purchasing effectiveness. This is an overlooked area of logistics and an area of real opportunity for many companies.

■ The last two columns (5.17 to 5.18) argue that "logistics is information" and that certain information systems will be essential to logistics success in the future. I believe that these new information systems capabilities are the key enabler of enhanced supply chain performance.

LOGISTICS AS A PROCESS

5.1	Logistics as a Process

December 1993

A growing number of companies are beginning to think of their activities as processes rather than a group of distinct tasks or functions. A few companies have even reorganized, using *key processes* as the unifying organizational principle. The "product development process" and the "account development and retention process" are as familiar to some companies as the old functional labels we used to use. This movement to "process thinking" opens new horizons for logistics in that it provides an avenue for bringing integrated logistics thinking into your company.

There always have been two schools of logistics. (This is why it has been so difficult to find a broadly understood, common definition of logistics.) One school defines logistics in a narrow, functional way—generally as traffic and distribution (T&D). T&D minimizes freight and warehousing costs, with negligible attention to the linkage to areas like customer service, inventory management, manufacturing, sales, forecasting, and marketing.

A second school views logistics as an "integrating function"—

linking all the activities in the supply chain. Prof. Bernard "Bud" LaLonde of Ohio State University uses the term "boundary-spanning activity" to refer to this type of logistics practice. In his definition, logistics is not a focused functional activity, but rather one that enables the integration of activities across functions. In this regard, logistics takes on a supply chain focus, whose success is predicated on the presence of an expert assessor of trade-offs, a skilled negotiator of policy, and an effective enforcer of the new operating protocol.

An effective way to promote this expanded role for logistics is to position logistics as a process, not as an activity or function. The logistics process spans the activities of many traditional functions, which must be coordinated in a harmonious way. Like all processes, logistics requires a "process owner," who designs and manages the process to operate in the most effective way.

I suggest that the following three important sub-processes are part of the logistics process:

- ***Integrated production and distribution strategy development***, which establishes the physical infrastructure of manufacturing and distribution assets (or capabilities if outsourced) that form the network of facilities from which products are sourced and flow to market. The scale, capacity, location, and operating methods (e.g., full stocking warehouse vs. cross-dock facility) form the infrastructure.
- ***The replenishment process***, which involves the materials-planning activities, including forecasting, inventory planning, production planning, and manufacturing scheduling.
- ***The order management process***, which spans the traditional customer service activities of order entry and order processing through inventory allocation, order fulfillment, picking, staging, transport, and delivery, through the financial activities of billing, collections, and deduction reconciliation.

In large part, this new view of logistics as a process has led to today's heightened interest in the discipline. Viewing logistics as a process will provide a firm platform for building a truly integrated business.

FORECASTING AND INVENTORY MANAGEMENT

5.2 Forecasting: An Overlooked Element of Logistics

July 1987

Forecasting is an activity that is not always associated with the logistics function. Nonetheless, a well-designed forecasting system can contribute significantly to logistics performance. It also can help to both lower inventory costs and improve customer service. Yet I have found that too many companies treat forecasting as an afterthought; they view forecasting as a necessary evil required for budget assembly once a year.

Basically, a sound forecasting system should address such questions as:

- Who should be responsible for forecasting? Sales? Marketing? Logistics? Some combination of the above?
- Are forecasts reviewed and agreed upon by key departments throughout the organization?
- What horizons and time periods should be used for both long-term and short-term forecasting?
- How are statistical and judgmental considerations combined?

If the forecasting process is inadequate, the various parts of the organization often end up operating with different numbers. Sales will create a forecast that reflects desired sales levels, logistics and manufacturing will modify these numbers to reflect their perception of reality, and finance will be operating with a third set of budgeted numbers. How can an organization achieve unity and harmony if each part of the company is working with a different set of numbers? Unfortunately, the answer is often that it cannot.

Choosing appropriate techniques is crucial to forecast performance. There are many fine reference works in this area, and an increasing array of commercial software packages are available. These packages include a variety of statistical and exponential smoothing models for short-term forecasting, regression and econometric models for longer term forecasting, and product life-cycle models for new product forecasts.

Additionally, several leading companies have exploited electronic linkages with their customers to improve their forecasting performance. These companies are linked electronically with their customers' retail warehouses (or in some cases, retail stores) in order to obtain data on current sales rates and pipeline inventory levels. This information assists these suppliers in understanding the *real demand* for their products.

Finally, a good forecasting system recognizes that forecasting is not an exact science. As one of my colleagues has often remarked, "Forecasting is easy, as long as it is not about the future." Nor is it a panacea that will resolve or eliminate logistics concerns. It is, however, an important element of logistics that can contribute to enhanced logistics performance.

5.3	**The Six Sins of Forecasting**

April 1991

My friend and former colleague Tom Gunn once quipped, "Forecasting is easy—as long as it's not about the future." As Tom's comment indicates, the future is uncertain and difficult to predict. Companies, therefore, face numerous challenges when developing a demand forecast.

For example, many consumer products companies are trying to operate with a 25- to 60-percent forecast error (on the stock-keeping unit level) in their one-month-out forecasts. This error range wreaks havoc with inventory levels and customer service performance. "Best practice" companies, on the other hand, consistently are able to achieve 15- to 20-percent forecast error rates.

Companies that perform poorly in their forecasting typically commit two or more of the "six sins of forecasting" detailed below:

1. ***Letting finances drive forecasts***—In too many cases, the demand forecast is dictated (or overly influenced) by financial considerations. Companies commit to a demand forecast that is not their best estimate of likely sales, but instead a conservative estimate that they are confident they can exceed. This practice allows managers to "make the numbers they commit to make."

   Recently, I saw a division of a large company forecast minimal sales for the third quarter and zero sales for the fourth for a hot product because it did not want to increase its profit commitments for the year. This practice sabotages inventory planning, customer service levels, and raw materials purchases. The correct practice is to develop a *true* operational forecast, that is, the best estimate of real demand. You may provide a range around this point to allow flexibility and appropriate conservatism in financial planning, but you need the best point estimate available if you are going to manage your operations properly.

2. ***Having no forecast "owner"***—Companies need a formal forecasting process and a specified forecast "owner." This person manages the forecasting process and tracks forecast performance. The forecast owner can be from any relevant department (such as sales, marketing, materials planning, or manufacturing), but must merge input from all of these groups when preparing the forecast.

3. ***Having insufficient analytical support***—A good forecasting software package and analytical "tool set" is critical for good results. Such tools let you predict the influences of trends, seasonality, and other factors. Although adjustments based on

sales inputs, market factors, competitors' actions, and the like are essential, they are not enough.

4. ***Using a single forecasting approach for everything***—Different approaches are required for long-term and short-term forecasting. The tools employed, the time horizons, the planning buckets, and the level of detail all should be different for each of these forecasts.

5. ***Having no SOPM***—Holding a sales and operations planning meeting (SOPM) is essential. This is a formal meeting (or conference call) among all key players, in which they review recommended forecasts and operating plans and then reach consensus on the best estimates and plan. The SOPM should include people from sales, marketing, product management, logistics, materials planning, and so forth. Depending on the environment, the SOPM may be conducted weekly, biweekly, or monthly.

6. ***Failing to track forecast error***—Best practice requires measurement and tracking of forecast errors. This identifies any bias and gives a scorecard of forecast effectiveness. Furthermore, error tracking helps determine inventory levels. Forecast error should always be reviewed during the SOPM.

It's easy to see how avoiding these "sins" can significantly improve forecasting performance. This is an area in which logistics managers have an opportunity to shine by taking the lead in driving excellence and serving as the owner of a key process for their companies.

5.4   Strategic Inventory Management

September 1989

Management of the corporate inventory investment has received widespread attention through the 1980s and will continue to be a focus of management scrutiny through the 1990s. Careful management of one's inventory investment is important for three reasons. First, stock availability is a key dimension of customer service. As customer service takes on increased importance as a competitive factor in virtually all industries, effective inventory management has become a key success element for many companies. Second, most companies have continued to emphasize the effective management of working capital, of which the inventory investment is a key component. And third, as companies focus on operating flexibility to respond to customers' needs, low inventory is essential. These competing demands for high availability, low working capital, and operating flexibility impose challenges and rewards for inventory managers.

Companies that have been most successful in meeting these challenges have addressed inventory on all three levels of the Strategic Inventory Management Pyramid (see Figure 5.1). The bottom two levels of the pyramid are commonly recognized as important factors in strategic inventory management, although many companies still have not developed these basic capabilities. The top level has received less attention by many companies and remains a major area of opportunity.

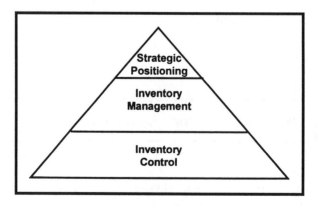

Figure 5.1   Strategic Inventory Management Pyramid

The base of the Strategic Inventory Management Pyramid, *inventory control*, focuses on building capabilities to achieve timely and accurate information on inventory status. Required capabilities include online, real-time information systems to record inventory transactions in a timely and accurate way, as well as the use of bar-coding technology and cycle-counting procedures to assure inventory accuracy.

The second level of the pyramid, *inventory management*, focuses on the use of effective forecasting techniques and processes, the use of the most appropriate planning rules and applications software for inventory management (for example, distribution requirements planning [drp]), and the collaboration across functions to achieve superior performance.

The top level of the pyramid, *strategic positioning* of inventory, involves a variety of initiatives to "change the rules of the game" and permit quantum improvements in inventory performance. This level is given less attention by most companies but offers the greatest potential benefits. Initiatives could include:

- *Inventory deployment*—Many companies are rethinking their inventory-positioning strategies. By inventory positioning I mean analytically determining what items are going to be stocked at what locations. The concept has been referred to as "prime/non-prime" and relates to the well-known principle of centralization of slow-moving items. However, new and emerging transportation options, lower transportation cost structures, and expanded merge or cross-dock capabilities have encouraged companies to take a new look at where they stock each item and have allowed many companies to considerably leverage their inventory investment.
- *Lead time management*—One of the best ways to leverage inventory performance is to shorten replenishment or inbound lead times. By reducing these lead times, companies can reduce their inventory investment significantly.

  I find it very interesting that in the automotive industry, most Japanese companies have focused their attention on inbound lead time management, concentrating on shortening the replenishment lead times for their after-market (or spare) parts. On the other hand, most U.S. automotive companies have focused their attention on the outbound side, rationalizing the stocking locations and the associated inventory deployment decisions. The benefits for companies that work to exploit both the inbound side (lead time management) and outbound side (inventory deployment) are enormous.

- **Channel integration**—Another tactic at the highest level of the pyramid involves channel integration. Achieving visibility in one's sales channel by working more closely with customers to understand actual demand patterns by item at the retail level (for example, by linkage to point-of-sale scanning data) allows a company to reduce demand uncertainty and the accompanying safety-stock requirements.

Companies must excel on all three levels of the Strategic Inventory Management Pyramid to achieve superior inventory performance. Those that do will achieve the benefit trifecta of lower investment and costs, higher service levels to customers, and increased operating flexibility and responsiveness.

5.5   Time to Consider DRP

September 1989

The operational advantages of distribution resource planning (DRP) have been well documented for many years. More and more companies are discovering that DRP systems can reduce costs, improve customer service, and better leverage their inventory investment. Yet a surprisingly large number of companies have not seriously considered the feasibility of DRP for their own operations. It is time for many of them to take a serious look at DRP.

Traditionally, most companies have used order-point logic to manage inventories. But order-point logic, particularly in multi-echelon warehousing systems, provides only limited visibility of actual demand at the regional warehouses. Inventory planners and production planners react to replenishment orders from a warehouse, not to actual usage and current system-wide inventory levels. This can lead to shortages when several warehouse orders arrive simultaneously or to excess inventory when sales slow down and production is not adjusted.

DRP systems provide a full view into the warehousing network by first examining demand at the end of the channel and accumulating requirements back through the warehouse network. This approach allows for full visibility of needs and better management of inventories.

It is important to keep in mind that DRP involves both inventory management and distribution planning. A module of distribution requirements planning (drp) extends the concepts of materials requirements planning into a multi-echelon-warehouse inventory environment. The results are time-phased replenishment schedules for moving inventories across the warehousing network.

In fact, DRP could be viewed as an extension of distribution requirements planning that is used to coordinate all logistics operations, including transportation, warehousing, manpower, equipment, and financial flows. DRP offers an accurate simulation of distribution operations with extended planning visibility, allowing logistics departments to manage all resources better.

In today's competitive environment, logistics continues to gain acceptance as a key strategic resource. Logistics managers with "vision" are searching for every opportunity to reduce costs or improve customer service and thereby create a distribution advantage. These continuing trends generate a heightened emphasis on planning and time-sensitive information on inventory and resources. DRP provides these capabilities while mobilizing managers to meet the challenges of a dynamic marketplace.

5.6   The Emergence of Channel DRP

June 1992

Over the past two years, numerous people have asked me the question, "Why haven't distribution resource planning (DRP) and distribution requirements planning (drp) been implemented in more companies?" After achieving wide visibility in the early '80s with numerous success stories documenting its considerable benefits (particularly with regard to order-point systems), drp has seemingly not realized its full potential.

Part of the reason may lie in the simplification of warehousing networks that has taken place over the past decade. DRP is most beneficial for multi-echelon distribution networks, networks in which one level of warehouses replenishes a satellite (or second) level of warehouses. Over the past ten years, as companies have re-evaluated their network strategies and reduced the number of warehouses in the network, the apparent need for and utility of DRP was perhaps diminished.

In the last two years, however, I have seen an increasing number of "channel DRP" systems, implementations that span echelons in the supply chain (see Figure 5.2). Basically, each customer distribution center (DC) is established as a stocking location in the manufacturer's DRP system. Information on daily withdrawals and current inventory levels is transmitted from the customer to the manufacturer at the end of each day. The manufacturer's DRP system manages replenishments from plants to both its own DCs and the customers' DCs as if the manufacturer owned the entire network. The benefits of DRP (lower inventories, lower transportation and operating costs, and better in-stock availability) often are considerable.

I recently have seen these channel DRP applications in the food sector, the medical products industry, and industrial products companies. In the medical products area, one major supplier has established the 27 warehouses of one of its wholesale distributors as nodes on its DRP system. Daily usage and forecast information is

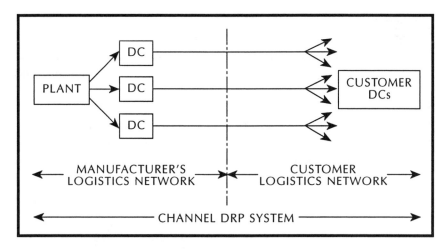

Figure 5.2 "Channel DRP" Systems

generated at these 27 locations, and inventory requirements and replenishment plans are netted back to the manufacturer's DCs, and ultimately to the materials requirements planning system, which runs the plants. Although still in the early stages of implementation, the system has reduced costs and inventory levels and improved service.

Channel DRP systems are beneficial for two reasons. First, given that supply chains from manufacturers to their customers are multi-echelon systems, a DRP replenishment system invariably produces superior results. Second, the logistics functions across the supply chain (from manufacturers to customers) traditionally are not connected. Channel DRP offers a way to build linkages among these functions (an arrangement often referred to as a quick-response, continuous replenishment, or partnership program).

Although only a few companies have implemented these channel DRP systems today, I expect them to be commonplace in five years. For example, one initiative under way in consumer packaged goods is an attempt to standardize interfaces in a system that limits point-of-sale scanning data and warehouse withdrawals into a channel DRP system. By developing a channel DRP system, your company can position itself to build both operating and strategic advantages.

5.7   Managing "Inventory in Motion"

August 1988

Over the past decade, we have witnessed profound changes in many aspects of logistics. None of the developments, however, has been as striking as the recent trend toward managing "inventory in motion"—that is, managing inventory while it is still in transit instead of waiting until it arrives at the warehouse.

By opting to manage "inventory in motion," a company essentially treats in-transit inventories as a working asset that can be closely tracked and controlled in order to enhance logistics performance. Consider the following examples:

- An electronics manufacturer carefully tracks an inbound container shipment of components from the Far East that is headed for its plant on the East Coast. When the container arrives at a West Coast port, a two-day supply of key components is stripped from the container and moved via air to the East Coast plant. The remainder of the shipment is hauled to the plant by truck.

- A replenishment shipment of consumer goods is sent by rail from the West Coast to a Chicago terminal. As the shipment reaches the terminal, its distribution is determined according to the current needs of each account. The shipment then is "cross-docked" and trucked to the appropriate customers; the goods are never placed in warehouse inventory.

The benefits to be gained from managing "inventory in motion" are significant. First, any inventory reduction has a double impact on return-on-investment (ROI) calculations: lower inventories increase profits (the numerator of an ROI calculation) *and* reduce the asset base (the denominator). This substantially improves ROI performance, particularly if inventories constitute 20 percent or more of a company's asset base. Second, the management of "inventory in motion" can considerably improve customer service performance by providing the flexibility to direct the inventory where it is most needed. Third, complications caused by obsolescence are markedly reduced under this system, because inventories are kept to a minimum.

The management of "inventory in motion" will increase in the future. To prepare for this development, logistics managers need to expand "real-time" inventory-control systems; employ new technologies such as EDI and satellite shipment tracking; place less emphasis on traditional warehousing; develop creative ways to utilize sortation, consolidation, pooling, and cross-dock facilities; and develop enhanced decision-support capabilities for shipment routing. The net result will be a distribution system that features better cost and service performance.

5.8 | Stages of Continuous Replenishment

August 1993

The practice of continuous replenishment is moving from a pioneering approach used only by leading-edge companies to a standard way of operating in many industries. Although it will take many years to reach that point, continuous replenishment is progressing from the embryonic to the fast-growth stage of its life cycle. Over the next three to five years, major manufacturers of consumer packaged goods and medical supplies, many categories of industrial suppliers, and others will have implemented continuous replenishment programs with most of their major customers. Many of the major manufacturers in these industry segments have continuous replenishment pilot programs in place today, and a good many of them have struggled with the decision of how best to structure these programs.

Before I discuss alternative ways to approach continuous replenishment, let me first clarify some key terms. As I use it, continuous replenishment refers to "vendor-managed inventory." That is, a manufacturer monitors inventory levels at its customer's warehouse(s) and assumes responsibility for replenishing that inventory to achieve specified inventory-turn targets and customer service levels. The manufacturer thus makes the replenishment decision, rather than waiting for the customer to reorder the product. The customer continues to own the inventory, but those inventories often are dramatically reduced in a continuous replenishment arrangement.

Continuous replenishment shifts the administrative burden for inventory management and replenishment at the customer's warehouse back to the manufacturer. This approach generally involves higher transportation costs because the manufacturer often must ship more frequently and/or expedite more shipments to achieve inventory-turn targets at the customer's warehouse. Continuous replenishment does, however, offer benefits to the manufacturer, such as greater visibility of actual usage (shipments from the customer's warehouse or even actual sales and sales forecasts at the customer level). When combined with a flexible manufacturing capability, this information theoretically allows the manufacturer to better manage its own inventory and production planning. In practice, manufacturers' inventories often must be increased to achieve customer service requirements.

Don't confuse continuous replenishment with *quick replenishment* (QR). Under a QR arrangement, the customer retains responsibility for monitoring inventory levels and for replenishment decisions. QR focuses on shortening the order cycle time (order-processing and delivery times). It also generally involves more frequent deliveries of smaller order quantities and thus can significantly reduce customer inventory levels.

Manufacturers have followed three different approaches for managing continuous replenishment programs:

- ■ *Customer-specified parameters*—The simplest approach for manufacturers is to ask customers to provide the maximum and minimum inventory reorder point parameters periodically as well as daily inventory balances. This is most efficiently done by fax or by an 852 EDI transaction. When the current inventory level falls below the established reorder point (the minimum), the manufacturer initiates a replenishment shipment and provides the customer with an advanced shipment notification (ASN-856 EDI transaction) so the customer can receive the product.

- *Full reorder-point control*—An alternative approach is for the manufacturer to maintain more active management of customer inventory levels, generally using classic reorder-point logic. The challenge here is the number of EDI transactions required to maintain active control of inventory at your customer's warehouse (see Figure 5.3). Although not all of these transactions are essential, companies generally require four or five of these transactions to operate effectively and often evolve so that they operate with most of these transaction sets.

- *Channel DRP*—This is the most comprehensive approach for continuous replenishment. It involves setting up each of your customers' warehouses as a node in your distribution resource planning (DRP) system and establishing a protocol to capture the key transaction sets required to actively manage inventories at those nodes.

Manufacturers have used different approaches to continuous replenishment. Some companies have used only customer-specified parameters. Others have started using the customer-specified parameters approach and evolved to either a reorder-point control or channel DRP arrangement. Some companies have jumped right to a channel DRP approach. How do you determine the best approach for your company? Ultimately, it depends on the depth of your commitment to continuous replenishment, the level of skills and resources available to dedicate to the program, and your customer service strategy.

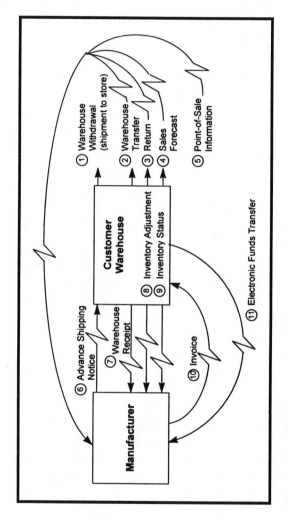

Figure 5.3   EDI Transaction Linkages

TRANSPORTATION AND WAREHOUSING

5.9	Achieving Transportation Efficiencies

October 1987

Most shippers have realized considerable reductions in their freight transportation bills in recent years. Deregulation (to varying degrees) across all modes of transportation, new technologies, and innovative management tools and methods have created cost-cutting opportunities for even the most unsophisticated shippers. Nonetheless, only a handful of companies have fully exploited transportation cost-reduction opportunities.

I have observed three stages in the development of a shipper's transportation management capability. Shippers in the first phase, or Stage I, have simply renegotiated discounts off selected freight rates. Generally, these shippers have not conducted a thorough analysis of freight movements, nor have they fundamentally altered their logistics flows.

Stage II shippers have fully analyzed their transportation movements and concentrated their volume with a limited number of carriers. They are obtaining the most competitive freight rates available, as well as benefiting from the administrative efficiencies gained by working with a smaller group of carriers.

The focus of Stage III shippers has extended beyond a concentration on just the freight bill to include attention to the full logistics costs. Along with concentrating volume with a limited number of carriers, techniques employed by Stage III shippers have included:

- Negotiation of service standards for each major traffic lane
- Introduction of EDI linkages with carriers for shipment advisories, receipt acknowledgment, shipment tracking, and billing
- Use of bar codes to reduce transaction costs further and improve inventory and document accuracy

- Use of a reliable transportation system as a critical element of a just-in-time scheduling and inventory-control system

Interestingly enough, the savings from reduced administrative costs, increased operating efficiency, and lower inventory realized by Stage III shippers generally equal or exceed the transportation savings. In other words, a Stage II shipper can again match the transportation cost savings it has already realized by becoming a Stage III shipper.

To reach Stage III, transportation managers must collaborate with managers representing other functions (MIS, inventory control, finance, etc.), as well as outside parties such as carriers. It is not surprising, then, that fewer than 10 percent of all shippers have arrived at Stage III. For the shippers that do reach this plateau, however, opportunities abound for reducing costs in every area of logistics.

5.10 Carrier Selection in the '90s

January 1991

Historically, companies in the business of supplying transportation services have assumed that shippers choose a mode and carrier based strictly on price. Although cost always will remain an important selection criterion or "buyer value," many shippers have changed (and are changing) their perspective. Today, service performance has become the predominant buying criterion for an increasing number of companies.

Put another way, shippers used to select the low-cost carrier. Period. Now, however, more and more shippers are screening carriers that meet their service performance criteria and only then negotiating price. A fundamental paradigm shift has occurred. This may not always be obvious to carriers because many shippers still negotiate aggressively, seeking the best bargain they can obtain.

Nonetheless, if you look beneath the surface, the primary buyer value has shifted—from price to service performance.

I believe that the carrier community at large (with some notable exceptions) has been slow to recognize this shift and thus has been slow to respond to the changing needs of shippers. Yet whether carriers acknowledge it or not, shippers are turning to total quality management and emphasizing customer service as never before. The result has been a renewed focus on low inventories, tight inventory management, and just-in-time manufacturing.

With smaller, more frequent shipments, the penalties for unreliable carrier service intensify. A late shipment can disrupt production schedules and impair customer service, causing significant cost penalties and customer dissatisfaction. Many companies, moreover, now realize that their delivery carrier is an important link to their customers, which frequently view that carrier as a representative of the shipper's company.

Therefore, along with the traditional carrier evaluation criteria such as price and financial stability, the new buyer values include delivery reliability, equipment condition, damage record, driver courtesy and appearance, information and systems capabilities, flexibility and responsiveness, ease of doing business, and a commitment to service excellence. These new values have (and will continue to have) an increasing influence on the carrier selection process.

In many ways, the new buyer values may appear unfair to carriers. Shippers want more for less—enhanced services at a low cost. But shippers are the customers, and those carriers that provide what the customer wants will get the business.

To get ahead of the curve and win, carriers must push their service performance to new heights. They must gear their operations to meet the new performance standards. How can they do this? The options include investing in market research to better understand the shipper's specific buyer values; enhancing delivery reliability; implementing a quality program to enhance performance in key areas; investing in information systems to provide better operating efficiency, control, and service performance; developing a "shipper-friendly" perspective, including simplified rate structures, and comprehensive shipment tracking; and tailoring service to individual customer needs.

5.11   Quality Programs for Transportation Excellence

November 1995

Within the past year, my firm, Andersen Consulting, conducted a comprehensive survey of our 1,200 shipper clients to determine the progress they have made in developing quality improvement programs for their transportation activities. Although the results are encouraging, they also point out the tremendous opportunities still available to many shippers. I would like to share some of the survey's findings and suggest directions for all shippers to consider in the management of their transportation programs and their carriers.

It is interesting to note that only 59 percent of the shippers surveyed have developed formal transportation quality improvement programs. Over the past decade, many noted practitioners and academics (most notably Dr. John Langley of the University of Tennessee) have articulated the value of the best approaches for achieving quality in transportation. Nonetheless, more than 40 percent of the shippers surveyed have not yet instituted a formal process for managing this activity.

The good news is that significant progress has been made in recent years. Of the 59 percent of shippers that have formal quality programs, more than three-fourths (76 percent) started these programs within the past three years. As the success of these programs grows and as this success is communicated at industry forums such as the recent Council of Logistics Management conference, I believe the number of shippers that have yet to institute quality programs will decline. Additionally, I expect to see shippers that do have active quality improvement programs enhance and expand them.

Clearly, there is room for improvement. For example, of the 59 percent of respondents that already have quality improvement programs:

- Only 61 percent have a carrier-certification program.
- Only 36 percent provide written confirmation of service standards to their carriers.
- Only 20 percent have computerized reports that measure and tract carriers' compliance with service standards.
- Only 19 percent measure the total cost of carriers' noncompliance with service standards.
- Only 20 percent provide formal recognition to top-quality carriers.

Furthermore, 16 percent of these shippers do not conduct formal performance reviews with carriers.

Customer service excellence is an increasingly critical competitive weapon in almost every industry—and quality transportation is essential if companies are to achieve that level of excellence. A quality improvement approach to transportation management does make a difference. Shippers that have not already done so need to initiate a quality improvement program, and all shippers should work to expand and enhance the scope of these programs.

5.12 Transactions vs. Relationship Buying of Transportation

January 1990

I find that there are basically two types of buyers of transportation services: transaction buyers and relationship buyers. Transaction buyers tend to move from deal to deal to obtain the desired services at the most economical rates. Relationship buyers, by contrast, concentrate on forming value-adding relationships with their carriers; that is, they try to create an environment where the whole exceeds the sum of the parts.

Transaction buyers, for their part, tend to be effective negotiators and deal-makers. They generally work in a more adversarial manner than relationship buyers. These are the buyers who sometimes schedule bid presentations back to back so that the departing carriers meet the next bidders in the lobby. They tend to focus on the short term, guided by an overriding objective of minimizing costs.

Relationship buyers endeavor to form long-term partnerships with their suppliers and work to identify operating relationships that benefit both parties. They invest the time needed to understand their carriers' current traffic lanes and attempt to provide movements that complement those patterns. They offer long-term contracts, encourage carriers to invest to lower costs, and share the benefits of cost reductions and operating improvements with their carriers. (Table 5.1 summarizes the differences between the two types of buyers.)

Which is the better buying approach? I think that each has its place. Relationship buying, when it can be achieved, is the preferred course. Yet in actual practice, it is not appropriate to all circumstances. Relationship buying, for example, becomes difficult when the shipper has seasonal or unpredictable loads. Some pro-

Attributes	Transaction Buyers	Relationship Buyers
Horizon	Short Term	Long Term
Objective	Minimum Cost	Maximum Value
Relationships	Adversarial or Neutral	Partnerships
Key Skills	Negotiating/Deal Making	Analytic/Problem Solving

Table 5.1   Comparison of Transaction and Relationship Buyers

viders of transportation services, moreover, do not have a propensity for relationship buying. It makes little sense for a shipper to push a relationship arrangement with a party that is not willing to participate. Under these circumstances, transaction buying is a better choice.

Most shippers predominantly follow either a relationship or transaction buying approach. I contend, however, that shippers must become effective at *both* types of buying. Both have their place, and both can work well for shippers. The trick is to develop the skills necessary for each as well as the ability to size up a situation and apply the appropriate approach. Mickey Mantle said that switch hitting added 25 points to his batting average. Effectively utilizing both buying methods can add points to a shipper's bottom line.

5.13 Reinventing the Warehouse

November 1994

In recent years, we have witnessed profound changes in how companies view their supply chains and the role distribution centers play in those supply chains. Three main factors are driving these changes. First, new technologies allow companies to operate in a fundamentally new way. Second, new transportation service offerings provide opportunities to redesign product flows. And finally, customer demands are shaping distribution patterns. Three of the most compelling of those demands are as follows:

- **Cost**—The pressure to reduce expenses to preserve shrinking margins and to respond to growing, global competitive forces.
- **Customer service**—The need to provide ever-higher levels of service to customers. This includes enhancing basic service performance in the areas of in-stock availability, on-time delivery, consistent delivery times, and errorless transactions. It also includes the need to provide value-added services to

customers as a source of competitive differentiation. These value-added services include rapid delivery (e.g., same-day or next-day service), continuous replenishment, drop shipments to stores or points of use, the building of mixed pallets, full EDI transaction capability, and so forth.

- *Speed*—The desire to compress the full supply chain pipeline, as a way to enhance responsiveness and to reduce assets and costs.

In response to these compelling requirements, a handful of innovative companies have begun reinventing the supply chain and changing the role of the warehouse within it. This new development has given rise to several radically different operating principles:

- *Flexible manufacturing*—Maximum flexibility is built into manufacturing by operating JIT/fast cycle manufacturing facilities with rapid changeover capabilities. This system allows a manufacturer to offer daily production runs of most items, contract manufacturing that provides flexible capacity, and the use of "mini-facilities" for both fabrication and assembly located close to the market.

- *Flow-through distribution*—The goal of a reinvented supply chain is to eliminate or limit the number of stocking locations. This implies fewer distribution centers, holding slower-moving items in a central location, and designing distribution centers with a cross-dock capability. This last strategy merges inbound shipments from different plants, from vendors, or from a master distribution center with locally stocked items into outbound shipments to customers in the local or regional market. The goal of many companies is to increase the volume of freight shipped in a flow-through manner.

- *Shifting of value-added activities to the distribution center*—Distribution centers are expanding into activities traditionally performed in the plant. These activities include packaging to order, kitting, minor assembly, labeling, and customized packaging to meet individual account requirements.

These emerging operating principles are changing the role of the traditional distribution center considerably. This "reinvented warehouse" can provide a powerful competitive advantage and position the logistics function as a key player in your company's success.

5.14 Should You Switch to "Cross-Dock"?

September 1990

Traditionally, distribution has tended to follow one of two patterns. Under the first option, *source-based* distribution, companies centralize inventory and move outbound shipments to customer locations, typically by less-than-truckload (LTL) or parcel delivery (although truckload may be used as well). Under *market-based* systems, by contrast, companies stock inventory locally, fill customer orders from local stock, and then ship them to nearby customers by the best delivery means.

Today, many companies are considering a third option—*cross-dock* operations—which, under many circumstances, combine the best features of source-based and market-based distribution. What follows is a discussion of the advantages and disadvantages of each option, with a suggested framework for evaluating cross-dock distribution for your own operations.

Source-based distribution is desirable when outbound shipments are truckload (TL) and/or when minimizing inventories is a key concern. This option doesn't work well, however, when short delivery lead times are required or when a considerable amount of freight moves via LTL and parcel carriers.

Market-based distribution is preferred when short delivery lead times are necessary and/or when inventory-carrying costs are less significant. However, high handling costs, unstable demand patterns, and high in-stock availability requirements tend to make this option less attractive.

Cross-dock operations, as noted in Figure 5.4, often combine the

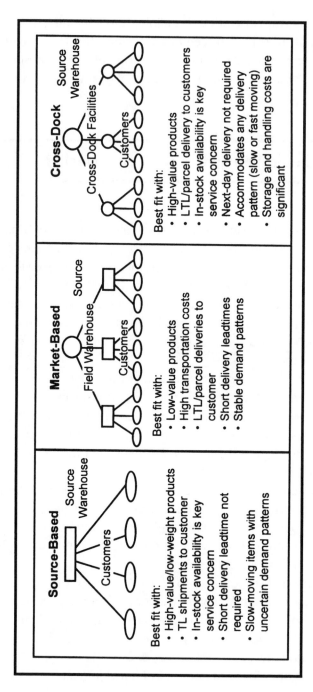

Figure 5.4   Advantage of Cross-Dock Warehouse Operations vs. Source-Based and Market-Based Options

best features of source-based and market-based distribution systems and, under many circumstances, provide better customer service at a lower overall cost. Cross-dock operations allow for leveraging of the inventory investment by avoiding multiple stocking locations. Cross-dock systems also eliminate the costly double handling of a product that occurs at market-based warehouses. Furthermore, they can eliminate the lengthy lead times of source-based LTL and parcel shipments by using speedy TL movements to break-bulk locations where products are redirected for local delivery.

Under many circumstances, the adoption of cross-dock distribution produces a quantum leap in logistics performance. Cross-dock operations, however, are not right for all circumstances and must be carefully evaluated. Additionally, because they substitute information for assets (inventory and warehouses), these systems require first-class information systems as well as advanced planning capabilities.

These strictures notwithstanding, many companies are finding cross-dock to be a way to gain both customer service and cost advantages. It is an option many companies should consider.

PURCHASING

| 5.15 | Purchasing Strategy for the '90s |

October 1990

Purchasing has not traditionally been considered part of logistics. Yet in an era in which most companies define logistics as "the management of the flow of materials from source to ultimate point of use," we clearly must include purchasing as an integral part of the process.

And it's an important part! For many companies, the cost of purchased goods and services accounts for more than 60 percent of the cost of goods sold. Effective purchasing management, therefore, is critical to effective operational performance.

Nonetheless, many companies have not focused management attention on enhancing this critical logistics function. They have maintained the same strategy, organization, operating methods, and information systems they have used for the past decade.

Other companies, however, have successfully used a powerful new framework to achieve quantum improvements in their purchasing effectiveness. I call it "Segmented Purchasing Strategies" (SPS). SPS manages purchasing activities not along traditional commodity lines but based on the nature of the risk and the economic opportunity potentially available from each product group (see Figure 5.5).

In the SPS framework, the term "exposure to market risk" refers to such factors as sourcing options, security of product availability, quality variances and importance, and supplier responsiveness. "Economic opportunity," on the other hand, refers to the potential to reduce costs or to achieve added value through service enhancements, innovations in features or characteristics of purchased goods, and/or streamlining of administrative procedures.

To plot materials on the framework, you segment each product family and assign the various segments to one of the four quadrants of the SPS to be managed in a distinctive way. They are:

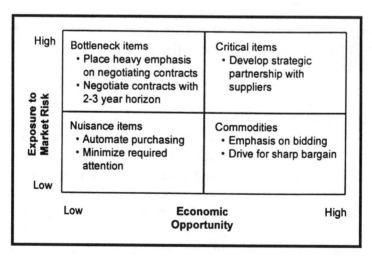

Figure 5.5   Segmented Purchasing Strategies Framework

- **Nuisance items**—These may include items such as office supplies and other low-dollar-expenditure/non-critical items. The strategy is to minimize the attention devoted to these items, preferably through mechanization of the purchasing process or through outsourcing.
- **Bottleneck items**—These important, low-volume materials are most effectively purchased through contracts. This locks in a secure supply, while freeing personnel to manage the high-impact items.
- **Commodities**—These low-risk/high-opportunity items are best purchased through a competitive bidding process. This allows companies to leverage their volumes for maximum cost advantage. For high-availability and low-relative-risk items, a purchaser can have suppliers compete for the business to obtain the best bargain.
- **Critical items**—Purchasers need to develop partnerships with suppliers of high-risk/high-impact items. Purchasers should focus key resources on securing, developing, and managing these partnership arrangements.

The world of purchasing is changing. The SPS provides a framework to redirect purchasing resources in a strategic way, which maximizes economic gain while minimizing risk.

5.16 Beyond Purchasing to Strategic Supplier Relationships

October 1995

When asked why he robbed banks, the infamous Willie Sutton once replied, "Because that's where the money is." Similarly, over the past five years, leading companies have discovered that "strategic supplier relationships" (SSR) are where the money is.

Strategic supplier relationships create significant economic value by synchronizing supplier and customer operations and by relentlessly driving waste and duplication from the system. SSR has uncovered huge savings and mutual benefits for those companies and for their partners, and they have driven these benefits directly to the bottom line. My colleague Paul Matthews points to the following recent successes:

- A U.S. packaged goods manufacturer increased its gross margin by 83 percent in two years by utilizing SSR.
- A U.K. retailer more than tripled its net income through SSR.

I want to caution that SSR extends far beyond approaches such as "purchasing excellence" and "supplier partnerships," which have received much media attention (and much criticism) in recent years. Though these concepts undoubtedly have created some benefits for the companies that employ them, they fall short of the potential available through SSR. "Purchasing excellence" too often degenerates into squeezing selected suppliers for a few extra bucks. "Partnership" approaches often fall apart because they tend to focus more on who gets the benefits and less on how to create real economic value.

Companies generally follow a three-stage process to implement SSR:

1. *Consolidation*—The first step is to give more volume to fewer suppliers, which drives down costs. This does not mean squeezing suppliers for lower prices; rather, it means shifting volume to more efficient suppliers and creating value through greater economies of scale. Improved quality is an important outcome of supplier consolidation. That's because SSR concentrates volume with suppliers that can produce low-cost, high-quality products and that can deliver according to prescribed schedules.

2. *Coordination*—In stage two, companies examine and synchronize the full range of supply chain activities. They integrate their operations, while eliminating duplication and waste in areas such as order processing, materials planning, inventory management, distribution, and transportation. This is the

point at which companies using SSR do supplier quality certifications, so that receiving can be streamlined and inbound inspection eliminated. They also shorten their delivery lead times as much as possible to reduce costs and enhance flexibility.

3. ***Cooperation***—The next stage of the SSR process is enhanced cooperation, particularly in product development, manufacturing, and logistics. Suppliers and customers work as an integrated team to leverage their combined knowledge. For example, together they can redesign component parts of a product in order to reduce production and assembly costs, they can use joint capabilities and knowledge to reduce manufacturing costs, and they can reduce logistics costs through reconfiguring product flows and leveraging their total transportation movements. Furthermore, they can closely coordinate new product introductions to minimize start-up costs and to assure a fast learning curve. These companies use the Japanese concept of "doing it right the first time" to create added value from day one.

Logistics managers are uniquely positioned to initiate and lead an SSR program. Because they represent the integrating function in a company, logistics managers have a view of the entire supply chain and therefore can work with suppliers to make the extended supply chain a source of competitive advantage and an economic engine for their companies.

INFORMATION TECHNOLOGY

5.17 Information and Logistics

May 1990

The dominant considerations for logistics managers have traditionally been warehousing, transportation, inventory management, and, of course, customer service. Our traditional focus has been on the "physical," with success dependent on the effective management of critical factors such as people, capacities, equipment, technologies, and operating methods.

Although these considerations will always remain important, the effective use of information to manage and control the logistics function has increased in significance of late. One manager puts it this way: "Logistics is becoming more information-dependent. In fact, I can almost say 'Logistics is information.' We could not meet today's cost and service demands with yesterday's information system."

Increasingly, the evidence suggests that he is right: Logistics is information. Competitive success and effective cost and customer service performance will be contingent on effective information systems capabilities. Having outstanding functional knowledge and working hard no longer will be enough. Effective information systems capabilities will be even more critical to superior logistics performance in the 1990s.

Information will continue to be substituted for all forms of assets. For example, in the inventory area, EDI linkages with suppliers, carriers, and customers as well as advanced applications software for better decision making will allow companies to reduce inventories significantly. Similarly, effective information systems will allow higher utilization of transportation equipment, thereby allowing companies to operate with less equipment—effectively replacing information with equipment.

Future logistics information systems must possess several capabilities. Systems must be able to process transactions rapidly and accurately. Real-time systems using accurate data-capture technologies will have to be incorporated in all functional areas, including warehouse operations, transportation, and inventory. Robust functional capabilities, tailored to a company's operations, also will be important. In addition, advanced decision-support capabilities will be needed to make better strategic, tactical, and operating decisions. Analytic capabilities such as logistics network modeling, transportation routing and scheduling, "real-time" production scheduling, and warehouse and inventory simulation must be available to logistics managers. Finally, information linkages with all trading partners—including vendors, customers, and carriers—will need to be developed.

Information systems will be an important differentiator of logistics performance in the future. Investment opportunities will need to be carefully evaluated, but generally those companies that invest in these information capabilities today will position themselves to gain future competitive advantages.

5.18 Leveraging Logistics through EDI

September 1987

Electronic data interchange (EDI) is providing logistics managers with new opportunities to leverage performance. Although the technology and standards have been available to most companies for some time, implementation of EDI has been slow. Nonetheless, more than a few progressive companies today are using EDI to obtain an operating edge.

Electronic data interchange involves the direct, computer-to-computer transmission of inter-company transactions. Although many people still think of EDI in connection with order transactions, documents exchanged electronically often include invoices, shipping advisories, debit or credit memos, and shipping papers. EDI, in fact, can be used to link a company to all external parties—not just customers, but suppliers, carriers, public warehouses, freight forwarders, and customs-house brokers as well.

The advantages of EDI, of course, are greater in relationships with a large number of transactions and with high-speed-delivery requirements. Nevertheless, progressive organizations of all sizes are finding that they can obtain logistics leverage from EDI in several ways, including the following:

- **The shortening of the order cycle time (by up to three days)—** Companies have used this lead time reduction in a number of ways. Some have reduced delivery times and thus significantly reduced inventory levels—both safety stocks and cycle stocks; others have used these three days to increase freight consolidation opportunities and thus reduce freight costs. Still other companies have employed a combination of these strategies.

- **A reduction of clerical labor for all transactions, particularly order processing—**A major supermarket chain, for example, reduced clerical labor by 75 percent.

- **A reduction in the error rate for orders, invoices, and other documents—**The well-proven rule of thumb that eight to ten times the effort is ultimately required to process an incorrect order or invoice than a correct one translates into a considerable labor savings from EDI.

- **Improved control of operations—**For example, up-to-date shipment status information can be obtained through EDI linkages to carriers.

In today's competitive environment, astute logistics managers are searching for every opportunity to reduce costs or improve customer service and, therefore, obtain a logistics advantage. As part of this effort, forceful managers are overcoming barriers to EDI implementation, such as customer passivity or technological timidity, and are forging electronic linkages with a variety of external parties to leverage their logistics performance.

6 TECHNIQUES FOR SUPPLY CHAIN EXCELLENCE

This chapter outlines a variety of analytical techniques and tools that can enhance management understanding and command of the supply chain. The techniques highlighted here offer practical ways to deepen insights and to support decision making.

The 14 columns in this chapter fall into three clusters:

- The first six selections focus on process analysis and the approach to redesign that analysis may find warranted. The process analyses discussed seek to enhance integration, simplification, quality, and communication, within and across organizations. "SOPM: Improving Internal Communications" (6.3—June 1989) outlines an enduring approach which has added substantial benefits to many companies. Also, note that "Re-engineering Logistics" (6.1—May 1991) is followed several years later by a column in Chapter 7 which boldly predicted the "End of Re-engineering" (7.17—December 1994). You may choose to read these columns in sequence.
- The next four columns (6.7 to 6.10) explore cost analysis, showing how new measures can more accurately capture the impact of logistics on corporate profitability. "Moving Beyond 'ABC' Analysis" (6.7—March 1994) provides a new, valuable twist on an old technique.

■ The last four selections (6.11 to 6.14) describe a handful of analytical tools that can strengthen logistics planning and management. "Tackling Unproductive Inventories" (6.11—November 1989) provides visibility to an underutilized approach for simple, yet very sophisticated, inventory analysis. Scores of companies have benefited from the approach, yet it remains an underexploited technique.

UNDERSTANDING THE SUPPLY CHAIN PROCESS

6.1   Re-engineering Logistics

May 1991

The concept of "business process re-engineering" is receiving heightened attention in business literature and among top executives in an increasing number of companies. Many companies, in fact, already have achieved notable success by re-engineering their business processes. Recently, for example, a major Japanese automobile manufacturer reduced its accounts-payable department by 50 percent by redefining the work processes. Similarly, a large, diversified U.S. company quadrupled its planned efficiency improvements in administrative functions through a business process re-engineering approach.

Business process engineering (BPE)—and, by extension, re-engineering—involves redesigning (generally simplifying) the work tasks performed to conduct any activity. It has several important premises:

■ BPE must focus on processes, not functions. For example, the order management process may involve numerous functional units (sales, customer service, credit, logistics), but it must be viewed as an integrated process. Similarly, the fulfillment process spans logistics, manufacturing, and finance activities.

- BPE generally begins with a "clean sheet" approach. One must operate outside the constraints of traditional thinking and not be bound by historical precedent. As such, the goal is quantum, not incremental, improvement. For example, one company has adapted a "10x" goal; employees are striving to improve operational efficiency tenfold.
- BPE tries to focus accountability and responsibility. Most activities should be performed within a focused cell, where a single individual or group can have responsibility for the entire task. This change empowers employees in new ways and enriches their jobs, as well as allowing for markedly higher efficiencies.
- BPE generally involves, but is not limited to, changes to a company's information systems. Information system enhancements are often an important consideration but are meaningless without substantial simplification of the business processes.
- Finally, BPE focuses on eliminating waste and improving the speed of operational and administrative activities. It forces a company to continually question what value is added in every task performed. If no value is added, the task is eliminated.

BPE has added efficiency to many logistics activities. For example, in the warehouse, incoming back-ordered items are flagged at receiving and are routed directly across the dock to shipping. The wasteful activities of put-away and let-down are eliminated for the back-ordered items. In the order-fulfillment area, orders are handled more carefully to reduce the number of returns. In the transportation area, shipment planning is integrated with carrier scheduling to allow more efficient operations.

By spanning functional boundaries, BPE promotes the integrated logistics concept. It provides a framework to focus on business processes, not compartmentalized functional activities. And finally, it provides a mechanism for logistics managers to drive productive change in their companies.

6.2 | Simplifying the Logistics Process

September 1988

Based on almost two decades of observing numerous companies work to improve their logistics systems, I have found that the most successful are those that focus on simplifying the underlying business process. I have seen too many companies try to automate or streamline transactional processes that are archaic, unnecessary, or poorly conceived.

There have been exceptions, of course—companies that realized the importance of first simplifying or improving the underlying process before overhauling the system. Consider the following examples:

- A company designing a new freight-bill auditing and payment system builds in a pre-audit capability that pays transportation charges upon material receipt. Thus, in one step, it eliminates the need for cumbersome matching of freight bills with receipt records and rate tables.
- A company installing a new inventory-control system does not replicate its traditional order-point logic but improves the system through distribution requirements planning logic.

In both of these cases, logistics system designers first analyzed how to improve the underlying process and then proceeded to make that process more efficient through automation.

There are a number of reasons why companies fail to first examine and improve the underlying business or transactional processes. Many times, the processes span several business functions, thus exempting any one function from full responsibility (and accountability) for the activity. Sometimes tradition stands in the way, and sometimes a narrow charter limits success ("Our task is to automate the order-entry process, not change how our company does business"). Fortunately, all of these narrow perspectives can be addressed by managers committed to real logistics system improvement.

I have found that it is worthwhile to ask four key questions when analyzing a logistics or business system. They are as follows:

- Why do we perform each task? What value is added by it?
- Why are tasks performed in the order they are? Can we alter the sequence of the processing steps to facilitate efficiency?
- Why are tasks performed by a particular group or individual? Could others perform this task?
- Is there a better way for the system to operate?

These questions deserve to be asked as part of an examination of the underlying assumptions upon which a company operates. By making these inquiries and analyzing the results, you can move toward building logistics capabilities that provide real competitive advantage.

6.3 SOPM: Improving Internal Communication

June 1989

In too many companies, inter-functional bickering is the rule rather than the exception. When operating problems arise, each function points a finger at someone else. Common complaints include the following: the sales force promises unrealistic delivery dates to secure customer orders, marketing creates unmanageable promotional programs, finance squeezes inventory to unworkable levels, and distribution incurs too much cost for too little service.

These complaints, even when made in jest, are usually symptomatic of ineffective operational coordination. If they continue unchecked, they can erode a company's competitive edge and may lead to a significant business decline. At best, this contentious situation limits a company's potential and creates a poor working environment.

The most effective mechanism a company can use to assure operational coordination is the periodic sales and operations planning meeting (SOPM). My experience in reviewing the planning processes and operating methods in scores of companies strongly suggests that companies that schedule SOPMs perform better than companies that don't.

The SOPM is a periodic meeting of key functions (sales, marketing, production, distribution, and finance) to review recent operating performance and future operating plans. Issues addressed include:

- A review of customer service, inventory, and operating cost performance for the recent period.
- A review of the market conditions and the sales forecast to assure that all functions are using the same set of numbers as the basis for planning. One of the situations I encounter most frequently is different functions operating with different forecasts!
- A review of the production and inventory plan for the period ahead. This review can occur on either a product family or a detailed stock-keeping unit (SKU) basis.

Specifics such as the frequency of the SOPM (weekly, biweekly, monthly), the planning horizon (four weeks, three months), and the level of detail (SKU, product family) must be determined by each individual company's needs. But regardless of the details, the SOPM will provide numerous benefits. These include improving inter-functional communication, bringing elusive problems to light, providing a framework for thoughtful management decisions, and focusing a company's efforts on improving its competitive posture rather than on internal finger-pointing. The SOPM is the fastest, least expensive, and most painless action a company can take to improve its performance. Try it, you'll like it!

6.4 Using Function-to-Function Meetings

January 1988

The principal point of contact between a company and its customers has always been (and will continue to be) its sales force. Increasingly, however, many companies have worked to broaden the points of contact with customers and to formalize these contacts. Today, in several manufacturing concerns, functions such as engineering, management information systems, finance and accounting, and, of course, distribution routinely meet with important customers to develop better ways of working together. Many of the most productive and beneficial changes, in fact, have resulted from the meetings with distribution personnel.

I refer to these inter-company, multi-functional contacts as "function-to-function" meetings. These meetings are generally held annually, and they serve several useful purposes. First, they provide a forum for suppliers and customers to resolve conflicts. Participants can discuss their specific problems and concerns, and they can collectively develop solutions.

Second, these function-to-function meetings provide an opportunity for suppliers and customers to better understand each other's operating needs and constraints. This personal understanding of how poor delivery performance affects a customer's operations generally sensitizes and motivates a supplier's personnel to provide the desired level of service.

Third, these meetings open an avenue of communication between suppliers and customers. Customers know whom to call when problems arise, so that minor problems can be eliminated before they escalate.

Fourth and most importantly, these meetings provide opportunities for companies to identify new and better ways of working together. These improvements can involve minor changes such as modifying the way a supplier combines orders and loads a truck in order to facilitate quicker processing at a customer's receiving dock or changing the labeling on secondary containers to facilitate product identification. Alternatively, these improvements can involve more fundamental changes, such as the introduction of bar coding or electronic data interchange linkages between companies or the change to an alternative transport mode.

These function-to-function meetings must be planned and managed properly so as not to deteriorate into bickering and finger-pointing. When conducted properly, however, they can be a force for productive change. Leading strategic thinkers have said that the most underutilized strategic force is the management of buyer and supplier relationships. Function-to-function meetings provide an avenue for participants to initiate operating changes that provide competitive advantages to both suppliers and customers.

6.5	Quality in Logistics

December 1987

The application of the "Quality Philosophy" or the "Total Quality Concept" (TQC) to the logistics function is accelerating. Though originally limited primarily to traditional manufacturing operations, the TQC approach has increasingly been applied to service businesses (such as transportation carriers) and administrative functions (including all aspects of logistics) in recent years.

Many individuals, such as Deming, Juran, Crosby, and others, have developed their own specific philosophies and approaches to quality management. These approaches generally involve several common principles, including the following:

- *Strive for continuous improvement*—An operation, function, or "management system" should set continuous improvement as its goal and continuously work to achieve better system performance.
- *Understand the full cost of poor quality*—A company must understand the full cost of non-conformance to standards. The cost of "doing it right the first time" is always consider-

ably less than the full cost of finding and correcting quality problems.

- *Focus on causes, not symptoms*—A company must find and correct the root causes of poor quality and not depend on inspection to eliminate individual errors. Do not "inspect in" quality, but rather alter the system that causes the errors to occur. For example, inspecting outgoing shipments is a costly way to assure order accuracy. Rather, find the sources of any picking or labeling errors and correct them through training, workload balancing, work simplification, automation of data entry, or other appropriate means.

- *Monitor system performance*—Use statistical tools to monitor the performance of the system to identify when the system is out of control.

- *Use a team approach*—Quality is not just a management or a quality control department problem, but a company-wide concern. Companies that are able to establish an environment that removes fear of change, encourages participation, rewards innovation, and involves all employees in problem solving are the leaders in quality.

To date, only a small percentage of U.S. companies have applied TQC principles to logistics. The logistics community's involvement in TQC, however, is about to move from the embryonic to the growth stage as many logistics departments initiate TQC programs in the next few years. I anticipate that quality-driven approaches will become an integral part of the discipline of logistics management and that in five to ten years, logistics groups in more than half of all U.S. companies will actively practice TQC.

6.6 The Power of Non-Competitive Benchmarking

March 1994

Changing environmental and competitive circumstances demand that companies periodically alter the way in which they conduct business. In particular, far-sighted organizations have found that they must be nimble in adapting new technologies or new methods to improve their operating performance. Japanese companies have learned this principle well; they have repeatedly outperformed their U.S. counterparts in instituting systems that strive for continuous improvement. In fact, I think that one of the most important competitive assets of many Japanese manufacturing companies is their ability to learn quickly and then implement needed changes.

In response to the need to update their business practices, many companies—both in the United States and abroad—have experimented with different "change agents." For example, a number of U.S. companies have adopted a "total quality philosophy"—often in the form of "quality programs" instituted throughout the organization. A few companies have successfully implemented these quality programs and fostered productive change. Many more companies, however, have organized and implemented these programs only to find that their efforts have not resulted in any major improvements.

Agents of change are by no means limited to quality programs, however. Several companies have recently discovered "non-competitive benchmarking" as an alternative approach to productivity improvement. Non-competitive benchmarking involves comparing the methods and costs of generic business functions (e.g., order entry, accounts-payable administration, and warehousing) with companies in non-competing businesses that are perceived to be superior at a particular function. Although the requirements of a particular function may not be exactly the same in non-competing companies, organizations using non-competitive benchmarking are discovering that they are able to identify ways to improve their systems, practices, and methods for most business functions, especially those involving logistics.

Non-competitive benchmarking is not easy to implement and it requires a time commitment from key managers and supervisors. However, as many companies are discovering, it provides a pathway to making *quantum* improvements in the service and cost performance of a variety of business functions. We will be seeing increased application of non-competitive benchmarking in many companies in the next few years, and the logistics function will continue to be one of the primary beneficiaries of this approach.

COST ANALYSIS AND MEASUREMENT

6.7 Moving Beyond "ABC" Analysis

March 1994

The "ABC" approach to inventory analysis has been a simple yet very effective technique for stratifying individual items (SKUs) into logical groupings for management. Companies have used this method to prioritize their inventory management focus, modify their stocking and control policies, and derive enhanced customer service from a lower inventory investment. More recently, however, selected companies have modified the traditional ABC approach and have achieved very interesting results.

In the traditional ABC analysis (see Table 6.1), a company ranks each item (SKU) based on its dollar sales volume. One invariably sees a Pareto curve, where the top 20 percent of the items account for approximately 80 percent of the sales. Companies use this ranking to categorize each item as an A, B, C, or D priority. They then develop different inventory-stocking policies, review policies, and control rules (such as safety-stock levels) for each item.

Essentially, that traditional approach focuses attention on the high-volume items. Companies that use it generally find that it improves inventory and customer service performance.

Although this method of analysis is valuable, some companies recently have modified it. In addition to sales volume, they also have used *gross profit dollars* of each item and *line-item order frequency* of each item to stratify inventory classification.

These additional criteria focus management attention on the most profitable and the most frequently ordered items. Accordingly, they affect important customer service measures such as "orders filled complete."

Classification	% of Items	% of Sales Revenue	Implied Inventory Management Rules
A	8%	50%	Stock at all locations; review frequently/manage with maximum attention; push for rapid replenishment cycles; consider daily production runs
B	12%	30%	Stock at most locations; review frequently/manage for rapid replenishment cycles and frequent production runs
C	25%	15%	Focus stocking locations; routine administrative review; weekly production runs
D	55%	5%	Stock centrally; minimal administrative attention; less frequent review and production runs

Table 6.1   Traditional ABC Inventory Analysis

To perform a modified ABC analysis, these innovative companies follow a five-step procedure:

1. Determine the desired criteria for inventory stratification.
2. Develop a ranked list of each SKU for each criterion. For example, if you selected all three criteria (sales volume, profit margin, and order frequency), you would develop three lists of SKUs, with each list ranked in order according to those criteria.
3. Consolidate rankings by weighting each criterion, multiplying the ranking of each item by the weighting for each criterion, summing these products across the criterion, and reranking these final sums.

    Most companies are weighting criteria equally (33 percent, 33 percent, 33 percent), while some are using weightings such as 50 percent for volume, 30 percent for profitability,

Item Number	Consolidated Ranking	% Sales Volume	Cumulative % Sales Volume	% Profit $	Cumulative % Profit $	% Line-Item Order Frequency	Cumulative % Line-Item Order Frequency
	1						
	2						
	3						
	etc.						

Table 6.2 SKU Stratification

and 20 percent for order frequency. You can experiment with different weightings to assess which approach is best for your company.

4. Profile the SKU stratification as shown in Table 6.2
5. Define A, B, C, and D classifications by selecting appropriate cutoff points for the A, B, C, and D strata, taking into consideration the cumulative percentage distribution for all three criteria.

This modified approach to ABC analysis broadens the focus beyond just sales volume considerations and allows a focus on often-overlooked issues such as item profitability (which affects overall profitability) and order frequency (which affects customer service performance). Furthermore, this approach allows a company to test different criteria, different weightings of these criteria, and ultimately the impact of alternative classification policies on all issues of strategic importance to the company—sales volume, profitability, customer service, and inventory investment.

6.8 | Know Your Logistics Costs

November 1990

As part of the recent emphasis on operational effectiveness, many companies are beginning to look at the full costs of doing business with each customer. By conducting a cost/profitability analysis, they are realizing numerous benefits, both for the logistics function and for the company as a whole. They also avoid potential pitfalls that are best described by the following adage: "We may be losing money on every customer, but we make it up in volume."

To determine the profitability of a particular customer or a particular order, the company must make a careful assessment of two key factors. The first factor, which pertains to the revenue side, entails a full accounting of all discounts, deals, allowances, and promotional support. The second component of a cost/profitability analysis, which involves mainly logistics costs, requires an under-standing of how handling and shipping costs relate to order size and delivery characteristics.

Companies can perform the analysis in two primary ways: They can profile costs by order size or they can profile costs by customer type or by specific customer. Both analyses often lead to surprising results. For example:

- Through a cost/profitability analysis, a major medical supply company discovered that it was losing money on more than 23 percent of its accounts. Furthermore, it found that 18 percent of its customers accounted for almost 83 percent of its profits. Based on this analysis, the company modified its pricing structure to improve the profitability of its money-losing orders. It also refocused its marketing and sales efforts to increase business with high-profitability accounts.

- In another example, the logistics manager of a consumer packaged goods company was questioned by top manage-ment on the company's increased logistics costs. Using the cost/profitability analysis, he was able to show that the cost increases could be attributed to reduced order sites, increased drop shipments, and specialized handling and pallet configu-ration requirements. He also was able to demonstrate that on a normalized basis, logistics costs had actually decreased! The increase in costs was due entirely to the expanded services.

These examples illustrate how the cost/profitability analysis benefits both logistics and the company as a whole. Once the company has a sense of which types of orders and customers are most profitable, it can adjust its pricing policies and services accordingly.

Logistics managers, for their part, can break out of the cost-minimization mode and establish their function as a value-adding operation. If they can calculate costs (and benefits) of value-added services and delivery innovations, these managers can explicitly link their group's activities to pricing and account-development strategies. As a result, this information ultimately empowers the logistics function to assume a larger role in planning the direction of the company as a whole.

6.9	Insights from Customer Profitability Analysis

February 1988

Many astute companies are placing less emphasis on being industry leaders in overall sales volume or overall market share, focusing instead on being the leaders in the "profitable customer" segments. That is, they are recognizing that the cost of serving all customers is not the same. Some customers are more expensive to serve and, therefore, are inherently less profitable than others.

In adopting this new approach, these companies first analyze the costs of serving different customer groups and then focus their marketing efforts to expand their market share of the most profitable ones. This explicit strategy allows these astute companies to achieve considerably higher profitability than their competitors.

The distribution manager in most companies is in the best position to initiate and lead a customer profitability analysis. Although the costs associated with serving one's customers involve sales, finance, and other functions, most of the critical costs that determine the profitability of specific customers involve the distribution function.

Key costs to consider when conducting a customer profitability analysis include the inventory-carrying costs and the transportation costs needed to support a customer's delivery service requirements. For example, customers that require short delivery lead times, higher order-fill rates (e.g., no back orders), or special delivery requirements are inherently more costly to serve. Other important costs to consider when conducting a customer profitability analysis include order-processing costs (which are influenced by both the frequency and the method of placing an order), sales costs (generally analyzed as a function of sales call frequency and duration), sales-support costs, costs of carrying accounts receivable, manufacturing costs, and warehousing operations costs.

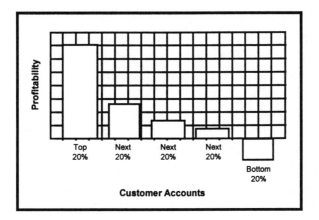

Figure 6.1 Sample Account Profitability Analysis

Top management in many companies often incorrectly assumes that all customers are equally profitable. These managers are often both surprised and enlightened by the results of a customer profitability analysis. These analyses frequently indicate considerable differences in customer profitability and can even reveal that some customers are inherently unprofitable (see Figure 6.1)!

Customer profitability analysis is useful in both developing a company's strategy and in designing the sales and operating policies to guide the business. Since top management and sales management in many companies have not considered this issue, the distribution manager has an opportunity to take a leading role in shaping company strategy and policy in this vital area.

6.10 Measuring Logistics Performance

April 1990

An important component of managing any business function—and particularly the logistics function—is performance measurement. Performance measures are used in many ways. They can set the standard that all employees strive to achieve, they provide feedback for employees on how they are performing, and they influence employees' behavior. Well-conceived measures can have a dramatic positive effect on logistics performance and on the interactions among individuals and departments within the logistics function.

I do not believe that there is one set of performance measures that is universally appropriate. Each company needs to design measures that are right for its particular strategy, operating environment, employees, and the specific needs of its customers. Too many companies either have not developed formal performance measures or have not given sufficient thought to the measures they use.

I would like to suggest four guidelines to assist you in developing performance measures. First, keep it simple. Identify easy-to-track and intuitively understandable variables.

Second, tailor the measures to what is controllable and important to the individuals or functions being measured. One company used "cases handled per labor hour" as the efficiency measure of its divisional logistics departments. Each division markedly improved its performance on this ratio, but the same cases were handled many times (both within a single warehouse and in trans-shipment between warehouses). As a result, these performance measures focused the company's managers on the wrong activities. Meanwhile, their competitors had developed effective cross-dock operations that reduced handling requirements and markedly lowered logistics costs.

The challenge in tailoring performance measures is to design them so they are controllable by an individual but do not sacrifice overall logistics performance for heightened performance in one particular area. For example, reducing transportation costs at the expense of overall logistics performance *is* no bargain.

Third, I suggest that logistics managers develop and use a variety of performance measures. Specifically, performance measures should be developed in the three categories shown in Table 6.3.

Fourth, performance measures should be assessed against a trend that occurs over time. Continuous improvement is the goal, and the

Customer Service Measures	Macro- Productivity Measures	Micro- Productivity Measures	
Percent line-item fill	Logistics costs as a percentage of sales	Warehousing cost per unit	Cost per mile
Percent order-fill complete: • At first pass • Within 3 days • Within 5 days	Transportation costs as a percentage of sales	Orders shipped vs. orders received Units picked per labor hour	Cost per ton-mile Cost per pound delivered
Delivery lead time	Aggregate inventory level	Percentage stock loss	Percentage empty miles
Order/shipping/ billing errors	Inventory time supply	Miles per vehicle per time period	

Table 6.3   Sample Performance Measures

graphic display of performance over time provides a useful management tool.

Designing effective performance measurements is not easy. But it is worth the effort, as effective measures contribute to improved performance, employee satisfaction, and harmony among all players.

ANALYTIC TECHNOLOGIES

6.11   Tackling Unproductive Inventories

November 1989

Keeping finished-goods inventory levels in balance with sales requirements is a key challenge for all companies. When inventory and sales are out of balance, a manufacturer faces two unpleasant choices. Either it must make painful and costly changes to its production schedules or its customer service levels suffer severely.

The key to effective inventory management is accurate forecasting (best achieved through shortened lead times and extended channel visibility) and sound inventory monitoring. A useful analytic tool for monitoring inventory levels is the Time-Supply Histogram. This diagram provides managers with a summary snapshot of item-level inventory performance and alerts them if item-level inventories are out of balance and immediate corrective action is needed. Using a histogram, a manager can assess item-level performance for a 100,000-item inventory from the information presented on a single page.

To construct a Time-Supply Histogram, compute the inventory

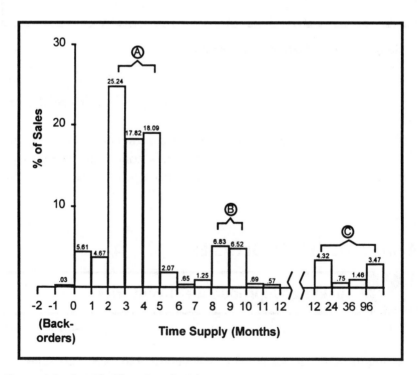

Figure 6.2   Sample Time-Supply Histogram

time supply for each item carried in inventory by dividing the annual sales of each item (at the current sales rate) by the current inventory level. Combine the items with similar time supplies (zero to one month, one to two months, two to three months, etc.) and add up the sales volume for each time-supply aggregation. The resulting chart might look like the one shown in Figure 6.2.

At this point, you can compare your chart with the "model" or desired histogram (Figure 6.3) to obtain a quick assessment of item-level inventory performance.

A review of the two charts indicates that the center of the primary mode on the sample histogram (point A) is 3 ½ months, about one month greater than the 2 ½-month mode of the "model" histogram. This indicates that the major items are somewhat out of balance and that adjustments may be needed. In addition, it appears that several other items (points B and C) deviate too much from the "model" histogram and deserve immediate attention.

The Time-Supply Histogram can provide the logistics manager with an early warning. Although more detailed and sophisticated analysis of inventories obviously must follow, this tool provides an overview of item-level inventory performance in a manner that is

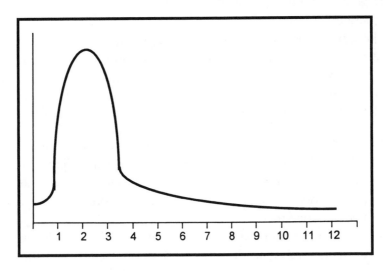

Figure 6.3 "Model" Histogram

quick to review, easy to comprehend, and focuses attention on critical items.

6.12   Using Logistics Models for Advantage

May 1988

Most companies operate in a complex environment today. Markets can quickly shift, sources of supply may be modified, and key cost components change. In addition, customers and product lines are added and deleted, sales volumes fluctuate, customer service requirements become more stringent, and responses to competitors' moves require changes to operating practices.

Facing such a dynamic operating environment, many logistics managers are turning to logistics-network modeling software (or decision-support tools) for help in redesigning their logistics systems. These advanced tools allow logistics managers to assess the cost and customer service implications of alternative logistics strategies and, consequently, place these managers in a position to exert greater influence over their companies' marketing and operating strategies.

Though the benefits of this technology may be clear-cut, many logistics managers and analysts have questions concerning the development or selection of network modeling software or analytical tools. In general, these managers would probably be well advised to do the following:

- Select modeling software with both optimization and "what if" simulation capabilities. These dual capabilities provide managers with a view of an idealized logistics system as well as an assessment of alternative logistics configurations specified by managers.
- Select modeling systems that can assess the full logistics network—inbound flows as well as outbound flows through all echelons of the distribution system.

■ Require that the modeling system use actual freight rates by shipping lane. Today, shipping costs are very "lane-specific." Models that use average freight rates or formula freight rates can yield misleading results.

■ Formulate the logistics system analysis in sufficient detail to yield meaningful results. For example, models that utilize an average shipping size rather than a profile of your company's actual shipment sizes can significantly mis-state transportation costs.

■ Utilize a modeling system that measures customer service performance (e.g., delivery lead time, anticipated order-fill rates, etc.) as well as total logistics costs.

A properly executed logistics system analysis, using a logistics network model, *is* an invaluable aid to logistics managers. With this tool, these managers can assess a range of alternative operating strategies and lead their companies in adapting their operations to the fast-changing business environment.

6.13 Leverage from Shipment Planning

November 1993

Virtually all industries are facing increasing customer demands for smaller, more frequent shipments. It is a challenge for distribution and transportation departments to provide this higher level of service while also limiting cost increases. Unless they have explicit strategies for responding to these new customer service demands, companies are seeing their freight and distribution expenditures increase by 5 to 15 percent or more.

Many companies, therefore, are pursuing a variety of strategies that will allow them to meet the demand for smaller, more frequent shipments at a reasonable cost. These strategies include rethinking their warehouse networks, changing their materials-handling methods, forming alliances with third parties, and considering alternative channel strategies (e.g., the increased use of wholesale distributors).

Another tactic that shippers are finding particularly effective is using sophisticated shipment-planning tools to help manage their transportation needs. Larry Lapide, developer of Andersen Consulting's SHIPMENT Planner® software, reports that "most companies can decrease their transportation bill by 5 to 20 percent through the use of a sophisticated tool to support mode selection and shipment consolidation planning. For a company with a large transportation bill," he says, "this capability can drop millions of dollars to the bottom line."

Effective shipment-planning software analyzes order profiles, consolidates shipments, selects the best shipment mode, and plans and schedules shipments. Here are some key capabilities such a tool often includes:

- *The ability to link directly to the open order file*—The planner can be run daily or weekly, but it should have the capability to consider information about orders that are outside the immediate planning window.
- *Analytic/optimization capability*—Specifically, the planner should have the ability to link inbound and outbound planning and to consider all modal selection options, alone or in combination, including truckload, inter-modal, multi-stop truckload, pool distribution, multi-stop with pool distribution, less than truckload, surface parcel, and air freight.
- *The ability to link up with the order-release and transportation-dispatching activities*—In particular, electronic data interchange links with carriers are powerful capabilities when tied to shipment-planning software.

An effective shipment-planning tool provides additional benefits beyond more informed mode selection. Specifically, it offers the user automated and faster shipment planning, better visibility of the opportunities for consolidation, and the capability to focus shipments with fewer, more efficient carriers. As customers' demands for smaller, more frequent shipments grow, a decision-support program that optimizes shipment planning will become a critical tool to help transportation managers keep their expenses in line.

6.14 | Opportunities in Routing and Scheduling

February 1990

Many companies that once relied on manual scheduling are discovering the profound benefits they can obtain by using routing and scheduling analytic software to dispatch their delivery fleets. These savings typically range from 5 to 10 percent of the total transportation expense, although they can exceed that by a considerable margin in a complex operating environment. Certainly, companies that have not considered routing and scheduling analytic software programs in recent years now should examine these productivity enhancers.

Historically, many routing and scheduling packages could not handle several real-world considerations—such as diversity of delivery-fleet equipment, mixed pickups and deliveries, backhauls, and unusual loading requirements—accurately or at a reasonable cost. That has all changed, however. The advent of sophisticated, economical computer routing and scheduling packages that can be readily adapted to an individual company's operating environment and address real-world problems provides new options for operating managers.

In fact, companies can use these newer packages as planning tools to examine a range of issues beyond fleet deployment, including fleet size, vehicle types, shift schedules, delivery frequency, depot location, and incremental costing. These broader applications, in turn, provide new savings opportunities.

As for which companies can benefit most from advanced routing technology, it appears that operations with the following characteristics generally enjoy greater cost-reduction potential:

- A large number of stops
- A large number of vehicles running concurrently
- A large number of stops per vehicle
- Uniform dispersion of stops over a service area

Yet just because a company has decided to adopt computer-based routing/scheduling technology, managers should not expect to see immediate results. Depending on the complexity of its operations, a company should plan on a two- to six-month development and implementation period to select the package, geocode the customer locations and road networks, build interfaces to the order-entry system, calibrate the software to the actual conditions, and train personnel. This long start-up period notwithstanding, many logistics managers are discovering that a computer-based routing and scheduling capability is one of the best investments they can make.

7 FUTURE TRENDS AND ISSUES: THE BROADER CONTEXT

While this chapter does not pretend to command a crystal ball, I can say that I am pleased with how these columns have provided insight about the future. Some of the predictions have come to pass, others are only now emerging, and a few remain mere glimmers on the supply chain management horizon.

Certainly the future is hard to predict. But the value of the effort lies in the insights gained into issues that are having or will have some impact on supply chain management and logistics. Whatever ultimate shape the impact assumes, management ignores these influences at its peril.

In many of my columns and particularly those in this chapter, I have also attempted to be a little provocative—to challenge an established point of view—with the goal of encouraging managers to think in a new way about these issues, to open their minds to "out of the box" thinking, and to encourage dialogue and debate about important issues. I think you will find that many of these columns encourage these goals.

The 20 columns in this chapter fall into two clusters:

- The first eight selections outline some emerging directions in logistics. From the historic opportunity to play a leadership role within your company to the need to rethink the consolidation trends of the past decade, these selections call a number of issues to management attention. "Historic Opportunity for Logistics" (7.1—June 1995) is intentionally excessive, with the aim of making clear that the world has changed and logistics has a new positioning in this world. My view is that logistics managers tend to be modest and less self-promoting than other functional managers (e.g., sales and marketing). This column is a call for action for logisticians and supply chain managers to take a more assertive position and exploit logistics as a strategic variable. "Back to Market-Based Warehousing" (7.3—October 1993) signaled (at an early stage) a fundamental shift in how logistics operates, while "Back to Rail" (7.5—June 1991) has not emerged as fully as predicted.

■ The next 12 selections (7.9 to 7.20) describe a variety of industry trends (such as Efficient Consumer Response in consumer packaged goods) and cross-industry trends (such as reverse-flow logistics) that will require revisiting many key aspects of supply chain management. "America Is Back, But Must Do More" (7.16—July 1990) predicted the re-emergence of U.S. competitiveness at a time when that direction was not clear and many were doubting the future of America.

As noted earlier, "The End of Re-engineering" (7.17—December 1994) boldly predicts the decline of this activity, while it appears to be at its peak, and "The Opportunity for Shared Services" (7.13—August 1995) also predicts an important future development. "The Emergence of Global Logistics" (7.18—October 1988) signals the arrival of a long-discussed trend, but predicts that it has arrived and will profoundly impact business and supply chain operations over the next decade.

The final column, "The Narcissistically Dysfunctional Organization" (7.20—August 1996), focuses on leadership and management styles and the impact of these styles on the performance of an organization and on the working environment.

EMERGING LOGISTICS DIRECTIONS

7.1	Historic Opportunity for Logistics

June 1995

It is a rare occurrence for an individual or a professional group to face the opportunity that lies before logistics professionals today. In fact, one can live a dozen lifetimes and never see such an opportunity. It is an opportunity created by fate, and one that every logistics professional should enjoy and exploit.

The opportunity before logistics professionals today is the chance to position logistics as a critical business function. In almost every industry, logistics has become a more critical activity than ever before. Logistics is at the heart of effective customer service, logistics is increasingly critical to effective cost and asset management, and logistics plays a key function in tying different activities in a company into a linked process, focused on efficiently serving a vital customer need.

There are several forces driving this historic opportunity for logistics:

- **Today's emphasis on customer service**—Business success in the '90s requires customer service excellence. That excellence is no longer an option or alternative strategy that companies can pursue, but rather a baseline capability required for success. Logistics is essential for effective customer service performance, for meeting both the basic needs of customers (such as achieving promised delivery dates, short order cycle times, and errorless picking) and the growing need for value-added services (such as continuous replenishment inventory management).

- **Transportation deregulation**—Though profound developments in transportation deregulation occurred 15 years ago and additional changes have taken place since that time, these developments continue to trigger momentous changes in logistics operations and performance. Transportation deregulation has triggered a reduction in rates and costs but, equally important, has triggered the development of new transportation services such as pool distribution, cross-docking, drop shipments to point of use, and expanded expedited delivery options. This has changed the economics of logistics, fundamentally altering the inventory and transportation cost trade-off. In essence, effective management of supply chain inventories becomes a more important factor.

- ■ *Process view*—As companies increasingly take a process view of their business activities, logistics' traditional role as an integrating function across corporate boundaries becomes more important. For example, logistics has demonstrated the skills needed to understand and bring together sales planning, customer service, order processing, warehouse operations, transportation, inventory management, and manufacturing planning into an integrated order-fulfillment process.

- ■ *Emphasis on asset management*—The increased importance of effective asset management, particularly working capital management, places emphasis on effective inventory management. Many companies that traditionally "turned inventory" three to six times per year now are achieving inventory turns in the double digits or higher. Logistics can place itself at the heart of effective inventory management, working across activities to raise customer service levels and reduce inventories.

Logistics managers should take heart in the historic opportunity before them. Circumstances have placed us in a unique position to take greater leadership within our companies. We can add value in ways unimaginable two decades ago.

I encourage you individually and collectively to seize this opportunity. We can add value to our customers and companies, we can achieve a greater level of fulfillment in our careers, and we can create a harmony and cohesiveness within our companies and across the channels we serve in a way never achieved before. Our destiny is before us—exploit it and enjoy it.

7.2   Future Directions in Logistics

August 1990

The future directions and trends in logistics management are taking shape somewhat differently from what was expected just five years ago. In response to these evolving directions, a number of companies have made fundamental changes to their logistics systems in order to position themselves to compete in the 1990s. Over the next few years, I expect many more companies to follow a similar path. What are these fundamental new directions? They include:

- *Fewer warehouses*—Companies are consolidating the number of warehouses used to serve their markets. In the mid-1980s, it was not unusual for a company to have 8 to 12 warehouses to serve the United States. Today, many companies have reduced their networks to three to six warehouses, while using transportation carriers that can provide fast and more reliable service. I predict that by 1995 almost all companies will serve their market areas with fewer warehouses.
- *More third-party services*—Very slowly, third-party logistics services are gaining ground. Although still only in the embryonic stage, third-party logistics services will expand rapidly in the 1990s. They gradually are becoming a more acceptable option for shippers, and by the close of the decade, they likely will be the option of choice. In the meantime, the range of activities performed by third parties has begun to

expand beyond warehousing and transportation to include broader functions.

- **New cost/service balance**—Clearly, customer service has become a much more important element of the marketing mix; as a result, the importance of the logistics function has grown. Companies now are systematically measuring customer service performance, and the service performance of the logistics system of the 1990s will be monitored as closely as logistics costs.

- **Globalization**—The movement toward global business operations, a trend that has been predicted for some time, will continue through the 1990s. Today, we see many logistics departments with a global span (not just a national focus and an export department). These global logistics departments manage warehousing, inventories, and transportation on a worldwide basis.

- **Channel integration**—The opportunity for quantum improvements in logistics performance in the 1990s will come from integrating one's logistics activities with those of suppliers and customers and then sharing the benefits. This channel integration has been referred to as "inter-corporate logistics," "quick response," or "supply chain management." Call it whatever you like, but watch for the leading companies in the 1990s to manage inventory, warehousing, and transportation across corporate boundaries.

- **Slower growth in logistics VP slots**—The broad acceptance of the concept of a logistics vice president will not happen in the 1990s. Most companies will consolidate more logistics activities under either the manufacturing, marketing, or materials-management function.

- **Expanded MIS role**—Logistics is becoming more information-intensive. Companies cannot be leading-edge logistics players without first-class information systems. Clearly, this requirement will continue through the 1990s. Information will continue to be substituted for assets (inventory, transportation, warehousing), and those companies with superior information systems will be able to improve customer service and reduce costs at the same time.

The 1990s will be an exciting and challenging time for logistics managers. More importance and more influence will be accorded the logistics function, but with that will come the responsibility to achieve higher levels of performance.

7.3   Back to Market-Based Warehousing

October 1993

Over the past decade, the logistics process in most companies and industries has been restructured significantly. In particular, most companies have undertaken a major consolidation of inventory stocking locations. Fifteen years ago, most consumer packaged goods companies operated with 10 to 15 stocking locations. Today, most have consolidated (or have plans to consolidate) to five to seven locations. Similarly, many industrial products companies now serve the United States from four or five locations, down from six to eight a decade ago. National pharmaceuticals and medical products distributors used to operate up to 90 locations; today, the three major distributors have between 45 and 48 facilities, with plans to consolidate further, to between 30 and 35 locations.

This consolidation of stocking locations over the past decade primarily was driven by two factors: transportation deregulation and enhanced information systems capabilities. Because of major transportation deregulation initiatives in 1980 (the Staggers Rail Act and the Motor Carrier Act), transportation costs decreased significantly. Even more importantly, new transportation services emerged, such as pool distribution, expedited delivery services, and so forth. These developments changed the inventory/transportation trade-off so that it became economical for companies to consolidate inventory locations and increase the service territories of their distribution centers. This logistics strategy allowed companies to operate with less inventory and, in many cases, to reduce their transportation costs as well.

At the same time, enhanced information systems capabilities made it possible for companies to plan and control inventory and transportation. For example, new systems gave shippers real-time visibility and control of inventory at all stocking locations. Furthermore, better decision-support systems allowed enhanced shipment consolidation planning. These expanded capabilities supported and encouraged the movement to consolidated stocking locations.

Although this forceful trend of consolidated stocking locations has significantly affected the entire supply chain over the past decade, I now am seeing the beginnings of a movement back to market-based warehousing. This new focus will encourage many companies to reverse their consolidated warehouse networks and expand their number of stocking locations.

This trend to more stocking locations is being driven by the accelerating customer service demands of almost all customer segments. As the marketing vice president of a major consumer products company puts it, "More stringent customer service requirements will drive manufacturers to use more stocking locations in order to have products positioned to replenish customers rapidly and in smaller delivery quantities."

As new services like continuous replenishment, vendor-managed inventories, shorter delivery times, and more frequent shipments (for example, changing from a schedule that calls for deliveries three times a week to daily deliveries) become commonplace, supplying companies (manufacturers and distributors) will be forced to expand stocking locations. Establishing additional distribution centers will become the only way these companies will be able to satisfy their customers' demands. Granted, they may find creative ways to manage these new locations (utilizing third parties and distribution alliances, for example), but the net result will be expanded warehousing networks and a reversal of the trend toward consolidated warehousing that has prevailed for the past 13 years.

7.4 | Future Directions in Transportation

October 1991

The transportation industry experienced profound changes in the 1980s. These changes were largely influenced by the major deregulation initiatives of 1980, the Staggers Rail Act and the Motor Carrier Act. Another important factor has been the increased operating sophistication made possible by advanced information systems.

There are still more changes on the horizon, however. Looking ahead, I expect to see the following developments as the transportation industry responds to a variety of forces:

- Core-carrier programs will accelerate the growth of large carriers. As shippers reduce the number of carriers they hire, it is the larger carriers that will be positioned to win a greater share of that business. Continued consolidation will mean that small, marginally profitable carriers will be squeezed even further and will eventually go out of business.

- There will be further concentration of the less-than-truckload trucking segment. The "Big Three" will gain a larger share of the market as they expand their service offerings in both long-haul and short-haul transport. In particular, investments to shorten lead times and make them more consistent will pay off with share gains.

- Dedicated contract carriage will continue to grow as shippers change how they buy transportation services.

- Smaller carriers that focus on specialized niche businesses will thrive. For example, smaller carriers may specialize in air-ride trailer or temperature-controlled services and thus will be able to achieve profitable growth.

- Increased use of rail and inter-modal services will replace over-the-road line-haul. This shift will be driven by environmental ("green") considerations, driver shortages, and enhanced rail service performance.

- Through inter-modal alliances or joint ventures, carriers will offer a broader set of services. These new ventures will include plans tailored to individual shippers.
- Shippers will continue to shift their emphasis from cost to service quality. Rather than buying transportation services based on the lowest cost offered by carriers that meet a minimum service level, shippers will buy based on service quality—as long as they can get that quality at a cost below a designated limit. This evolution will require carriers to become increasingly service conscious.
- New freight lanes will evolve. In particular, the evolution of a North American trading region will increase movements to and from Canada and Mexico. Also, recycling will create large, new freight flows. Foresighted carriers are positioning themselves to serve these new markets.
- Effective information systems capabilities will be even more important to carriers in the 1990s. These capabilities include decision-support tools to plan operations, as well as shipment tracking, service performance measurement, linkages to shippers through EDI, and cost control.

The 1990s will continue to be a turbulent time for carriers, offering both opportunities and risks. Carriers with insight into the future directions of transportation can position themselves to benefit from these developments. Those that do not face an uncertain future.

7.5   Back to Rail

June 1991

The 1980s witnessed a decrease in modal share for rail—at least where many manufactured and processed goods were concerned (see Figure 7.1). As shippers focused on inventory performance, they looked for consistent and short lead times from carriers, and

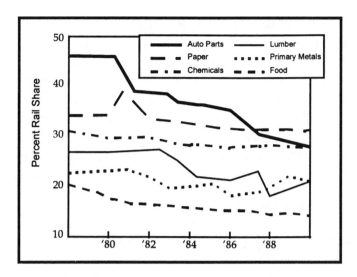

Figure 7.1 Railroad Market Share of Key Traffic Segments (Source: TRANSEARCH/Reebie Associates)

rail often did not measure up. Shippers also were won over by the "service sensitivity" shown by motor carriers. Many truckers listened to their customers, demonstrated flexibility and a willingness to tailor their operations to shippers' needs, and continuously focused on high quality and on meeting their customers' expectations.

In the past year, however, I have seen many shippers rethinking their modal choices and considering at least a partial return to rail. This reconsideration is being prompted by several factors:

- Several railroads have invested to improve the consistency of their lead times. These efforts are beginning to pay off, and service performance is becoming more predictable.
- Productivity advances, particularly in inter-modal, have improved both the efficiency and speed of rail service. Improved terminal operations, enhanced computer systems, and new technologies such as bulk-transfer facilities, integrated warehousing, and double-stack trains have allowed rail to be more competitive on both cost and service.

- The threat of stricter environmental regulations has encouraged shippers to seek an alternative to motor carriage. Potential legislation to limit over-the-road freight movements into California, the increased likelihood of restricting trucks in urban settings during specified hours, and more stringent air-pollution control and traffic restrictions all make motor freight a less attractive alternative.
- Furthermore, the growing driver shortage and fuel escalation clauses add uncertainty to the motor-freight pricing structure.

In spite of these developments, there has not been, nor will there be, a major shift to rail. But that doesn't mean that rail cannot reverse the declining trend in its share of manufactured and processed goods traffic.

To capitalize on this opportunity, railroads must focus on several objectives. First, they must invest in technologies that allow them to make their lead times more consistent, as well as shorter. Second, and most importantly, railroads must become more customer-focused—not in word but in deed! Customer-focused does not mean assigning an account manager to a customer; it means investing time to understand customers' real needs and then being sufficiently flexible to tailor one's services to those needs. Third, railroads must continue to invest in information systems that allow them to plan, monitor, and control operations better and that provide current and accurate shipment status information. Finally, railroads must assure quality through training, management development, and comprehensive performance measurement.

7.6   Competing as a Third-Party Logistics Company

December 1991

A major change has been occurring gradually over the past decade, and I see an acceleration of this development as we go further into the 1990s. Transportation companies and public warehousing companies are becoming *logistics* companies. This movement is formidable and irreversible because it is customer-driven.

As more transportation and warehousing companies rush to become logistics service companies (or, as they are commonly called, third-party logistics service companies), the competitive dynamics of this relatively new industry will change. Those third-party logistics companies that are well positioned will thrive; those that are poorly positioned will be required to compete on price and will have to accept lower profit margins.

What will it take for a third-party logistics service company to be positioned effectively in the 1990s? The answer, of course, is company-specific, but most providers will find they must offer at least some of the following:

- **Superior service quality**—The ability to provide differentiated, consistently superior customer service performance will be important. The hurdle for acceptable performance will continue to rise, and companies will have to show continual improvement to remain significantly above the base requirements. Furthermore, an emerging buying criterion will be whether or not a company has a quality program in place. Establishing such a program assures a continual improvement in service quality.

- **Strong information systems capabilities**—Leading-edge information systems capabilities will be a key requirement for third-party logistics companies that want to compete in the 1990s. These systems must encompass broad functionality in all areas of logistics (order management, inventory management, transportation, warehousing), full EDI capabilities, and the skill to customize systems to meet customer requirements and to link to customers' systems.

- **Full service**—The successful third-party provider of the future will undoubtedly demonstrate the capability to provide a full range of logistics services. This would likely include a knowledge of and a working relationship with transportation carriers (truckload, less than truckload, rail, inter-modal, air, and ocean) and with other service providers (those that handle customs clearance, import/export documentation, freight bill auditing and payment, and so forth). Furthermore, the capability to provide value-added services such as light assembly, packaging, kitting, ticketing, building mixed pallets, and the like also will be required.

- ■ ***Strong analytic skills***—Additionally, third-party logistics companies must consider providing their clients with the tools they need to analyze new operating options and to assess the cost and the customer service implications of those options. The skill to conceptualize forward-thinking, yet practical, operating techniques and strategies that enhance your customers' competitiveness will be an important attribute for third-party service providers.

- ■ ***Laser-like focus***—Successful third-party providers will be focused on a business or set of businesses where they can offer distinctive service strengths. This may involve an industry segment, specific handling characteristics (e.g., frozen cargo), or a set of value-added services (e.g., light assembly) where third-party companies can build a distinctive expertise.

The 1990s will be a decade of opportunity and challenge for third-party logistics companies. The movement of the market from the embryonic to the fast-growth phase will provide unlimited opportunities for providers that are well positioned to exploit them.

7.7	Getting Organized for the Late '90s

January 1994

Organizing a company's logistics activities is always a great challenge. In particular, many companies struggle with how to achieve the concept of integrated logistics (the coordination of logistics with other functions, such as sales, marketing, and manufacturing).

Beyond that, businesses also must sort out and coordinate cross-business logistics activities. This problem is particularly acute in multi-divisional companies, defined here as those that have separate divisions (generally organized around product categories) serving similar markets. In essence, the multi-divisional company has four organizational choices, which can be characterized as follows:

- ■ *Non-integrated*—Operating functions within each division (sales, manufacturing, logistics, etc.) operate separately in an uncoordinated way. In addition, each division operates independently, not leveraging logistics activities across divisions.
- ■ *Functionally integrated*—This type of setup features close coordination and integration across operating functions. Each division, however, operates independently.
- ■ *Divisional integrated*—In this arrangement, designed to maximize cross-divisional synergy, logistics is coordinated across divisions but remains unintegrated with other functions.
- ■ *Fully integrated*—In this case, there is close coordination and integration across operating functions and across divisions.

Within those general company organizational schemes, there are different ways of organizing the logistics function. The four diagrams in Figure 7.2 outline the logistics organizational schemes available for multi-divisional companies. What follows is a brief description of each:

- ■ *Option 1*, which features separate logistics operations, has been the traditional model for many companies. It has become less and less effective as customers demand higher levels of service and lower costs. This model coordinates poorly across functions, missing the synergies between divisions.
- ■ *Option 2*, which is characterized by a centralized logistics capability, has allowed companies to capture synergies across divisions while still preserving their autonomy. However, as customer requirements extend beyond combined shipments to a fuller set of value-added services, this model may fall short.

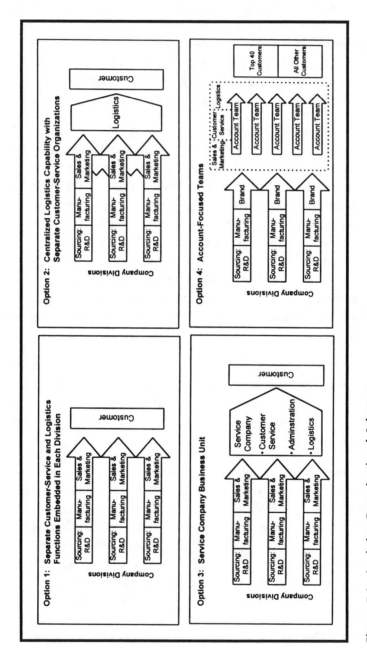

Figure 7.2 Logistics Organizational Schemes

- **Option 3**, which adds the "service company" business unit to the mix, allows full integration of all "account facing" activities while preserving the focus and management control gained from product-oriented divisions. This organization model is gaining favor in many companies.

- **Option 4**, which could be called the account-focused organization, may be the best model for the late 1990s. It is a bold way of connecting with the consolidating customer base that is developing in many industries. Several pioneering companies already are aggressively pursuing this model.

I have not seen a logistics organization that is right for all companies at all times. Market needs, customers, competitive requirements, and other enabling capabilities (such as information systems) all change. Organization of a company's logistics activities, therefore, must be dynamic, continually adjusting to these changing forces. In that light, these four options provide organizational models that companies can adopt as their logistics organizations evolve.

7.8 Logistics and the World of Virtual Retailing

March 1996

Of the many forces that are profoundly affecting almost all Americans and changing the way we live, two of the most powerful at work today are time and technology. With an increasing number of families managed by either a single parent or two working parents, time becomes a critical variable for most households to manage. In fact, any service or capability that can help households free up some time is highly valued—and will be even more so in the future.

At the same time, technology is beginning to penetrate everyone's life. Not only are people being exposed to technology at an earlier age, but they also are becoming more comfortable using technology. One need only visit a bank to see tellers sitting unoccupied while long lines queue up at the ATM machines to realize that technology is rapidly being accepted by people in all walks of life. In addition, more and more homes have (and are using) personal computers. As a result, online services are growing at astronomical rates.

One trend that is motivated by these two forces—known as "consumer-direct" or "virtual retailing"—is beginning to take hold. In a nutshell, it involves consumers selecting and ordering goods online rather than by visiting a retail establishment. The goods then are either delivered direct to the consumer's home or are made available for pickup.

Though consumer-direct is unlikely to ever dominate the retail scene, most analysts believe it will be a significant (and profitable) industry segment. Some estimates show it gaining as much as 15 to 20 percent of grocery sales in the next seven years.

Consumer-direct requires two key capabilities. The first is a virtual shopping and order management interface. This can be as simple as a catalog and a fax or as sophisticated as a "virtual" supermarket with aisles displayed on a personal computer. Shoppers can "walk" the aisles, select merchandise, obtain nutrition or other information about the product, and place and pay for their order online. At companies like Peapod in Chicago, thousands of customers are using such computerized ordering and product display interfaces today.

The second capability that is required for consumer-direct service is an efficient delivery mechanism. There are several approaches that are considered feasible:

- **Peapod approach**—A third party walks the supermarket aisles, assembling the order and then delivering it directly to a consumer's home.
- **"Value-added pickup point" approach**—A distributor establishes an item-picking capability in its warehouse, picks orders, and then delivers them to a pickup point (such as a gas station) where consumers can drive in and pick up their groceries on their way home from work.

- **"In-home delivery" approach**—Used by Streamline of Boston, which assembles grocery orders in its distribution center for delivery to consumers who have a three-temperature unit (shelves, refrigerator, freezer) at home. Streamline now has broadened its services to include laundry, film processing, and video rentals.

Although the efficiency and desirability of consumer-direct still must be proved, this development will have profound implications for logistics managers. It will be their responsibility to understand emerging consumer preferences and needs and then to develop new channels as well as new handling and delivery capabilities geared toward serving the individual consumer.

INDUSTRY AND RELATED DEVELOPMENTS

7.9　The Changing Role of the Distributor

February 1994

The wholesale distribution business has long played a vital role in distribution channel strategies in a variety of industries. Distributors traditionally have linked buyers and sellers, creating economic value by providing an efficient means of both selling and delivering products to customers. Nevertheless, a number of forces are challenging distributors to rethink their place in the supply chain.

Traditionally, distributors have added value through efficient performance of two primary activities: selling and physical distribution. By selling efficiently, distributors have provided a mechanism for smaller manufacturers to reach the market and for all manufacturers to reach more remote markets. In many instances, the key activities of new product introduction, in-store merchandising, in-office sales support, order taking and processing, customer service, training, returns processing, and problem solving were best handled by someone who understood the customer's perspective. Distributors filled the bill: They were close by and could frequently visit the customer. Furthermore, where relationship selling was important, distributors often were best positioned to build the strongest ties.

Similarly, on the physical distribution side, the distributor traditionally has been able to provide significant efficiencies and superior customer service performance. Valuable services have included local stocking, consolidated shipments with the orders from many suppliers, shorter delivery times, and value-added capabilities that could be done locally.

Some suppliers have opted to bypass distributors and go direct to the customer. In some industries, it is a smart strategy; in others, suppliers have hurt their market positions by losing the "distributor advantage."

Although the value traditionally provided by distributors is undeniable, their future role is more uncertain. Distributors are being affected by a variety of forces, including the following developments:

- Consolidation of both suppliers and customers has reduced the value provided by distributors. Where a Wal-Mart, for example, replaces ten small stores, direct distribution from a supplier direct to the large retailer becomes more feasible.
- Transportation deregulation has reduced transportation costs and greatly enhanced service offerings (such as cross-dock, pool distribution, and truckload with stop-offs). These new services and cost structures allow suppliers to reach many customers either directly or from a more remote location.
- Technological advances have challenged—or even pre-empted—traditional distributor activities. If customers can enter orders via electronic data interchange, the distributor may no longer be needed to take orders. Furthermore, freight consolidation software and materials-handling advances are challenging the need for local stocking.
- The number of companies with a supply chain perspective has grown. This management philosophy places tremendous

emphasis on careful inventory management while reducing the number of stocking locations and the number of times material is handled.

On the other hand, some countervailing trends such as shorter order cycle times, more frequent shipments, and continuous replenishment arrangements may reinforce the distributors' position.

Given these dynamics, what is the future for wholesale distributors and how should they respond to this changing environment? The only sure answer is that the distributors' role will most certainly change. How it changes will depend on a number of factors: the specific market served, the degree of customer consolidation, the degree of supplier consolidation, and the degree of growth of diffuse markets.

All distributors will need to continually reassess the evolution of their particular markets to identify where they can add the most value. They must invest in leading-edge information system capabilities so they can link seamlessly with other players in the channel, they must relentlessly drive costs out of their businesses, and they will need to aggressively search for ways to provide new services that add value to the supply chain. Some distributors inevitably will falter, but I believe there will still be a place for many distributors in the channel structures of the 21st century economy.

7.10 Winning Strategies for Distributors

February 1996

Wholesale distributors have had to navigate a treacherous course over the past two decades. In light of the significant consolidation of both supplier and customer bases that has affected most wholesale distributors, many observers have predicted the demise of this key player in the distribution channel. However—to paraphrase Mark Twain's famous remark—reports of the death of the distributor have been greatly exaggerated. In many industry segments, in fact, the distributor is thriving and is playing an even more critical role than ever.

Distributors add value in two key ways: They provide physical distribution efficiencies, and they provide both efficient and effective selling, market development, and account servicing. To build a continuing competitive advantage, distributors must leverage their expertise in these two areas, which are the foundation of their value as service providers.

My colleague Vic Orler has described how distributors can maximize their value to their customers by focusing on these two areas. In his research report for the National-American Wholesale Grocers Association titled "Wholesale Food Distribution—Today and Tomorrow," he characterized successful distributors of the future as being *network optimizers* and *market maximizers*. I believe these profiles can apply to wholesale distributors in virtually all industry sectors, and they can provide distributors with a powerful source of competitive advantage.

- The **network optimizer** plays a critical role in the channel by designing and managing product flows to minimize total channel costs. The network optimizer of the future will go beyond the traditional responsibilities of operating efficient distribution centers and consolidating outbound shipments to assess options such as plant-direct delivery, cross-docking, mixed pallets, drop shipments, and other strategies that can minimize channel costs.

  Network optimizers will proactively guide channel members (including suppliers and end users) to help them reduce supply chain costs. Furthermore, they will encourage and facilitate information sharing and channel management approaches in order to enhance customer service and reduce inventories and costs throughout the channel. These distributors will assume the role of "channel captain," earning their margins and adding substantial value as a proactive manager of the channel, rather than as just a provider of distribution services.

■ As **market maximizers**, wholesale distributors of the future will invent ways to help their customers enhance the effectiveness of their marketing and operations. Such value-added services can take many forms; which form they take will be limited only by the creativity of the distributor.

Some of the ways in which distributors can contribute to their customers' success include:

■ Assisting specific customers with information technology enhancements to improve channel effectiveness
■ Providing analytical tools or databases to enhance customers' understanding of their markets
■ Developing creative marketing programs for a particular industry
■ Providing simple tools to customers to assist them in disaggregating costs and understanding the true economics of servicing different marketing segments
■ Providing a menu of different services and programs from which customers may choose

Distributors are positioned in most instances to extend their reach and influence in the channel. In doing so, they can reinforce their competitive position, while enhancing their own and their customers' financial performance.

7.11   Efficient Consumer Response

July 1993

Efficient Consumer Response (ECR) has become the code word for how the consumer packaged goods channel will operate in the future. Essentially, ECR involves collaboration between consumer goods manufacturers and retailers to enhance merchandising effectiveness, inventory and material-flow efficiency, and the channel's administrative efficiency. ECR is expected to eliminate billions of dollars in waste annually from the consumer packaged goods channel. A few packaged goods manufacturers and retailers already have recognized this opportunity and have made impressive progress. However, most companies have been slow to grasp ECR, and opportunities remain enormous.

An analysis of the consumer packaged goods supply chain indicates that there are significant opportunities for improvement. Consider the following:

- Channel inventory in consumer packaged goods in the United States (the sum of manufacturers', distributors', and retailers' inventories) averages 84 days' supply. Various analyses have indicated that channel inventories can be reduced by up to 50 percent in some cases, freeing up significant working capital while enhancing product freshness.
- The administrative burden for ordering, order confirmation, invoicing, reconciliation, and deduction processing is substantial for most consumer products manufacturers and retailers. Deduction processing and reconciliation alone represent a multi-billion-dollar opportunity for the industry.
- The planning and management of the product flow can benefit from integrated approaches by manufacturers and retailers. Traditional channel distribution networks were set up to minimize transportation costs by maximizing truckload shipments. Modern approaches eliminate unnecessary storage locations, streamline product flows, reduce unnecessary handling, and utilize cross-dock and pool-distribution options where appropriate.
- By focusing on category management, analyzing point-of-sale data to gain marketing insights, micro-marketing at the store level, customizing promotions, and measuring promotional effectiveness, consumer packaged goods manufacturers and retailers have begun to utilize their joint resources to drive more store and product volume and to increase share.

ECR involves developing capabilities in a number of areas, as shown in Figure 7.3.

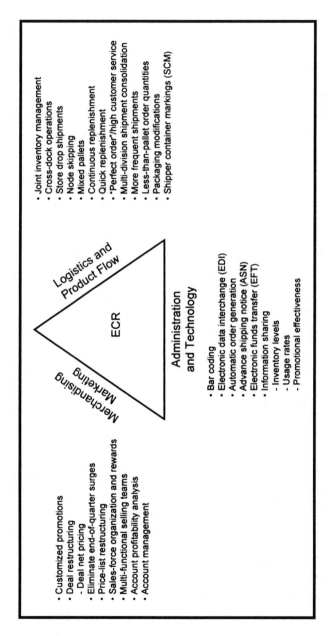

Figure 7.3   Key Capabilities for ECR

Effective logistics strategies are essential for successful implementation of ECR programs. In many companies, logistics managers are playing a leading role in developing and implementing the ECR initiatives and are critical to the success of ECR.

7.12 | The Emergence of the Distribution Utility

May 1993

The pressure on suppliers across virtually every industry today is unrelenting. Increasingly, powerful customers are demanding "more for less"—continuously higher service levels at lower costs. The higher service requirements involve a number of dimensions but invariably include smaller, more frequent shipments.

Many manufacturers have been innovative in developing efficient ways to provide smaller, more frequent shipments. They have initiated effective consolidation programs, utilized pool-distribution transport options, and restructured their logistics networks. As their customers continue to demand more, however, these manufacturers are at a loss as to how to squeeze "more service" (smaller, more frequent shipments) from their logistics operations. Typically, they encounter the following constraints:

- Less-than-truckload shipments will not meet the short lead time requirements.
- Manufacturers do not have sufficient volume to provide more frequent pool-distribution shipments.
- Manufacturers cannot economically justify more stocking locations closer to each major market.

An emerging alternative for companies is a "Distribution Utility," effectively a distribution alliance with manufacturers of non-competing products serving the same end customers. The "Distribution Utility" permits more frequent shipments to customers of the consolidated loads from all manufacturers in the alliance.

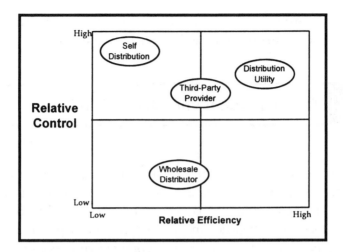

Figure 7.4 Control vs. Efficiency

A "Distribution Utility" is fundamentally different from a whole-sale distributor or third-party distribution company (see Figure 7.4). With wholesale distributors, for example, companies find that they give up significant control over their goods when they transfer products to these intermediaries. Also, this approach involves a second echelon of stocking, with the manufacturer generally being a first stocking echelon. Using a third-party distributor, on the other hand, gives a company greater direct control, but does not necessarily provide the opportunity for additional consolidations.

The "Distribution Utility" offers greater opportunities for consolidations, with only modest loss of direct control. However, "Distribution Utilities" are complex to set up and, like all alliances, are challenging to make work. Issues regarding cost allocation, service standards, operating policies, and product compatibility all must be negotiated. Furthermore, the acquisition of competing product lines by alliance partners can undermine a "Distribution Utility."

In spite of these challenges, more companies are seriously exploring the use of a "Distribution Utility" in order to provide their customers with "more service at less cost." As we approach the year

2000, the "Distribution Utility" will likely become a standard operating option for many mid-sized to small manufacturers.

7.13   The Opportunity for Shared Services

August 1995

The relentless pressure on almost every company to enhance customer service and at the same time reduce costs seems never-ending. Companies continue to ratchet up their demands for "more for less." Just when logistics managers have taken performance to new heights, customer demands and competitive pressures require them to do even more.

A number of multi-divisional companies have responded to this continuing pressure by exploring the concept of "shared services." This concept uses a shared corporate infrastructure for selected activities, which generally fall into two categories: administrative services and customer services.

Many multi-divisional companies have found that both the effectiveness and the efficiency of administrative functions such as human resources, finance and accounting, information systems, and procurement can be improved by centralizing their activities. These centralized "shared administrative services" provide tailored services that meet the needs of individual divisions while operating at a reduced cost. To accomplish that, they use advanced information technology and an understanding of the best practices of leading-edge companies.

Of particular relevance to the logistics function is the growing interest among multi-divisional companies in shared "logistics services" or "customer services" divisions. In essence, these centralized divisions carry out activities such as order processing, customer service, order fulfillment, distribution and transportation, and invoicing for all company divisions. They also plan and manage programs that deliver additional value for their customers or reduce customers' costs.

The growing presence of these customer services divisions is spurred by several factors:

- ■ ***The desire of an increasing number of customers to have a single point of interface with multi-divisional suppliers***—Customers increasingly want one shipment, one invoice, and one payment in order to simplify their operations and reduce their costs. Furthermore, as customers consolidate suppliers by eliminating smaller vendors, multi-divisional companies have an increased incentive to provide a single face to the customer.

- ■ ***The opportunity for multi-divisional companies to reduce costs for both themselves and their customers***—Shared services allow these companies to cut costs in such areas as order processing and customer service (by offering greater scale), warehouse operations (by combining facilities), and transportation (by consolidating shipments).

- ■ ***The opportunity to provide value-added services as a result of combining divisions***—Because of its increased scale, a shared customer services division is in a better position to implement value-added services that build operating linkages with customers. Examples of such services include electronic data interchange, continuous inventory replenishment, quick replenishment, pallet programs, and drop-shipment programs. These services often form the heart of "partnership" relationships that suppliers develop with their customers. As those customers continue to consolidate and become more powerful, it becomes essential for the manufacturers that serve them to offer such value-added programs.

The design and implementation of a services division is a challenging undertaking. It involves significant change and investment; it requires rethinking key processes such as order fulfillment and replenishment; and it requires working through difficult issues such as inventory ownership, organizational design, functional responsibilities, and performance metrics. Creating a services division, however, is worth the effort, as it can catapult a company's cost and service performance to a new level while enhancing the role, contribution, and importance of logistics professionals.

7.14 | Managing Reverse-Flow Logistics

August 1992

The past decade has seen a broadening of the concept of *logistics* to include supply chain management—the management of the flow of materials from source to point of use. Recognizing that total channel economics are linked, progressive companies today have expanded the logistics purview beyond physical distribution to include such areas as:

- Sourcing, purchasing, and supplier management and integration
- Manufacturing strategy, including decisions involving plant focus, product assignment, flexibility, quick changeovers, and just-in-time principles
- Inventory management, with a focus on speed of throughput
- Inbound and outbound transportation
- Warehousing, with consideration to opportunities for cross-dock operations
- Customer service, with attention to understanding the different needs of distinct segments of customers; at the same time, these suppliers are using logistics to identify value-added services that are desirable to those customer segments

In the years ahead, we will see a continued evolution of the logistics concept, with more attention paid to reverse-flow logistics—the flow of materials from the point of use to an earlier point in the supply chain (see Figure 7.5). Although it has been a common practice for some time in field service operations in telecommunications, computers, and other industries, reverse-flow logistics

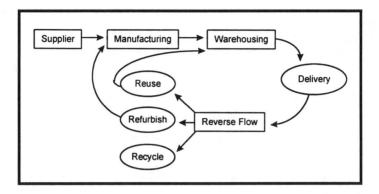

Figure 7.5　Reverse-Flow Logistics

will receive greater attention over the next decade because of growing environmental considerations.

Specifically, reverse-flow logistics adds three potential movements to supply chain logistics to accommodate the following:

- **Reuse of products or containers**—For example, the trend in the chemical industry away from drums to refillable iso-containers will dramatically increase reverse flows.
- **Refurbishing of products**—Products such as vending machines, computers, telephone equipment, and circuit boards are repaired and placed in inventory for resale. Many companies are moving to hold more unrepaired equipment in inventory and refurbishing to order in a just-in-time fashion.
- **Disposal**—Increasingly, both products and packaging are being recycled or returned for proper disposal.

Reverse-flow logistics requires different management approaches. Among the challenges are rethinking logistics networks, establishing distinct material- and inventory-control systems, and recognizing how the channel economics differ from traditional supply chain flows. In short, logistics managers will need to develop new tools and frameworks to plan for the increased reverse-flow logistics movements expected in the 1990s.

7.15 The Greening of Logistics

September 1992

Although environmental issues have been around for some time, the environmental movement will become more important and influential over the next decade. This development will have profound implications for logistics managers, as most aspects of our logistics operations will have to be modified (or rethought) in response to expanding environmental concerns. Over the next decade, environmental rhetoric will change to action, and companies will need to prepare for these new realities.

I see at least four areas where logistics managers will have to fundamentally rethink their operations:

- *Packaging and unitized loads*—Perhaps the greatest area of opportunity (and the area that historically has received the least attention) is packaging and unitized loads. New regulations and changing economics caused by escalating disposal costs mean companies will need to rethink their packaging designs. Reusable packaging, reusable containers, primary and secondary packaging with lower disposal volume, and collapsible racking that dramatically reduces packaging requirements will become standard for many companies.

  Moreover, organizations will have to maintain a higher level of packaging engineering skills and incorporate these skills into an integrated logistics organization.

- *Reverse-flow logistics*—Logistics networks will need to be redesigned with reverse flows in mind. Instead of disposing of non-consumable products, companies will recycle them for reuse. Recycling of packaging material will extend beyond the familiar cans and bottles to a broad range of materials. Moreover, the disposal of packaging will become the manufacturer's responsibility.

  The Council of Logistics Management is planning a research project in 1993 on reverse-flow logistics. This research should provide useful insights on how companies should rethink their logistics networks in light of these new operating requirements.

- ■ *Modal choice*—Shippers will have to shift more of their freight to rail and truck-on-freight-car (TOFC). This movement will be driven by regulations that limit trucking access to urban areas and by growing road congestion, which results in longer and less predictable lead times.

  This modal shift will not be easy for most companies. Many will have to contend with lack of easy access to rail spurs, greater attention to damage-resistant packaging, and adjustment to a less "JIT-friendly" form of transport. Although there will be resistance, this shift to rail is inevitable.

- ■ *Compliance*—Regulations will emerge in a haphazard fashion. Different countries and even various states and provinces will develop regulations at different paces. Therefore, companies will need to track environmental legislation impacting logistics operations.

Many of the situations outlined in this column may not happen until the year 2000 or beyond. Some will happen earlier. They *all* will have a profound impact on logistics operations.

It is not too soon to prepare for tomorrow. The era of "Green Logistics" will be here sooner than you think.

7.16   America Is Back, But Must Do More

July 1990

The 1980s were a disheartening period for many U.S. manufacturers. Jobs, markets, and pride were lost—or at least diminished—as foreign competitors proved to be more adept at designing, manufacturing, and distributing high-quality, low-cost products. Economists were debating if our wealth could be sustained through a "service economy," while many U.S. companies seemed to make negligible progress in their efforts to improve their manufacturing and distribution operations.

However, the 1980s closed on an optimistic note as a number of U.S. manufacturers sent notice that they were back. Companies such as Motorola, Westinghouse, General Electric, and Hewlett Packard have learned to compete on the basis of operational excellence. These companies have simplified operations, shortened lead times, eliminated waste, improved quality, and integrated all operational activities to lower costs, improve product quality, and win back markets. The results have been impressive.

Yet U.S. companies must do still more. Those U.S. companies that have not advanced over the past decade must act immediately; time is running out for them. Those U.S. companies that have advanced and have learned to compete on the basis of operational excellence know that their foreign-based competitors have not stood still—and will not stand still in the future. U.S. manufacturing companies must continue to improve and use every means possible to enhance operational performance.

The prospects of long-term success for U.S. manufacturing companies, however, also depend on issues beyond these companies' control. The United States must create an environment in which its manufacturers can compete. Specifically it must:

- ■ *Rebuild the physical infrastructure*—The United States must improve and extend its road, bridge, rail, and port networks so that manufactured goods can safely and economically be moved to both domestic and foreign markets. The transportation-infrastructure has decayed to the point where it can be a significant inhibitor to economic progress. This infrastructure must be both rebuilt and extended.
- ■ *Invest in educational excellence*—The quality of schools and the education provided in the United States is uneven and in many areas can only be described as deplorable. The economic, social, and national security costs of this failure are enormous. U.S. businesses spend vast sums to train workers in reading and basic mathematics. The United States cannot

be a leading economic power and world-class competitor without a highly trained workforce. Investments in education will pay for themselves many times over. Curtailing these investments is shortsighted and diminishes the long-term quality of life.

■ *Create level playing fields*—The United States is just beginning to pay for the excesses of the 1980s. The twin deficits (federal budget and trade) have created a false prosperity. The bills are coming due.

The federal budget deficit must be reduced to lower the cost of capital to U.S. companies. Companies cannot pay a premium of five or six points on the cost of capital and compete effectively on a global basis.

The U.S. trade deficit must be narrowed. The primary focus should be non-tariff barriers limiting access to the Japanese and other markets. The United States is asking many of its manufacturers to compete with one hand tied behind their backs. This is unfair and has been tolerated for too long.

Long-term competitive success for U.S. manufacturing companies is dependent on factors both within and beyond their control. All avenues must be aggressively pursued to assure prosperity in the 1990s and beyond.

7.17 The End of Re-engineering

December 1994

Re-engineering has been a powerful force in many corporate environments. Companies around the world have restructured their work processes to eliminate non-value-added activities, shorten cycle times, and enhance the effectiveness of their operations. In many cases, re-engineering has allowed those companies to reduce costs and increase near-term profits significantly. Importantly, many companies also have realigned their cost structures to allow them to compete more effectively in an increasingly global imperative arena.

At the same time, re-engineering has had its human costs. As thousands of middle-management and clerical jobs have been slashed, individuals have had to struggle with pain and uncertainty as these restructurings eliminated their jobs.

The good news is that, in the United States, we are seeing the beginning of the end of re-engineering. This sea change is good news on two levels. On the corporate level, many U.S. companies have shed excess costs and now are globally very competitive. On the personal level, the pain of adjusting to these economic dislocations will diminish.

While considerable re-engineering will continue over the next few years, its popularity as *the* approach to business improvement will decline. I believe that within the next three years, re-engineering in the United States will approach the usefulness of the experience curve, the analytic framework that drove many U.S. companies' strategic decisions in the 1970s. Similarly, re-engineering will remain a valuable technique, but managers no longer will depend on re-engineering to drive their businesses. (Note that Europe is behind the United States in this regard and will experience considerable re-engineering activity over the next five years.)

The bad news is that many managers and many companies are ill-prepared to move beyond re-engineering. These companies have developed an inward-looking culture. Their perspective is on cutting costs, enhancing short-term ROI performance, and achieving best-in-class benchmarks. Their perspective is not forward-looking—that is, focused on innovation, leapfrogging competition, investing for long-term competitive advantages, and finding new ways to do business.

In their *Harvard Business Review* article "Competing for the Future," Gary Hamel and C.K. Prahalad have framed the issues well. They suggest scoring your company on the issues in Figure 7.6 as a way to assess the degree to which it is forward-looking. Hamel and Prahalad argue that forward-looking companies will fall mostly on the right side in their responses to the accompanying questions.

How does senior management's point of view about the future compare with that of your competitors?

Conventional .... Distinctive and
and reactive   far-sighted

Which business issue absorbs more senior management attention?

Reengineering .... Regenerating
core processes   core strategies

How do competitors view your company?

Mostly as a .... Mostly as a
rule-maker   rule-follower

What is your company's strength?

Operational .... Innovation
efficiency   and growth

What is the focus of your company's advantage-building efforts?

Mostly .... Mostly getting
catching up   out in front

What has set your transformation agenda?

Our competitors .... Our foresight

Do you spend the bulk of your time as a maintenance engineer preserving the status quo or as an architect designing the future?

Mostly as .... Mostly as
an engineer   an architect

Figure 7.6   Hamel/Prahalad Scorecard (Adapted from "Competing for the Future," *Harvard Business Review*, July–August 1994)

Logistics managers can contribute to their companies' forward-looking perspective. They need to look outward to understand how markets and customer service requirements are changing and how innovative logistics solutions can help meet those needs in innovative ways. For example, years ago, innovative approaches such as the extensive use of third-party logistics providers and "flow-through logistics operations" (which rely on flexible manufacturing, cross-dock distribution facilities, and continuous replenishment through channel inventory management) were beyond the imagination of many companies.

Over the past five years, however, a number of companies have used such logistics innovations to build significant competitive advantages in cost and customer service performance. Logistics managers can help lead their companies beyond cost cutting and streamlining to leadership through innovation in their operations. To do so, they must explore how innovative approaches can create a fundamentally new cost/customer service paradigm for their companies.

7.18 The Emergence of Global Logistics

October 1988

Since about 1980, much has been written about global markets and the global corporation. Assertions that worldwide markets are becoming increasingly homogeneous and that future market segmentation strategies will span geographic boundaries have garnered attention—and sparked much debate.

Certainly this movement to global markets would have profound implications for a company's operations function. No longer would that group organize and manage production and distribution on a national or regional basis. Instead, it would design products for worldwide markets and manage a worldwide production and distribution system that would source, produce, and deliver products to customers in a way that best met overall cost and customer service objectives.

Within the last year, I have observed many companies taking the global logistics concept beyond the talk stage. In fact, I have seen the emergence of global corporations and global logistics functions across a number of industries. In several companies, logistics is no longer separately organized into U.S. and international (or export) areas. Rather, a single broad-based logistics function develops global sourcing strategies, global production strategies, and global distribution strategies, managing the worldwide distribution function much as most U.S. companies manage national distribution today.

Needless to say, the management of a global distribution function and the management of both material and information flows in a global logistics system present additional challenges for distribution managers. Sourcing and warehousing strategies require an understanding of a broader set of legal, tax, and regulatory requirements; inventory strategies must consider packaging, labeling, and language differences; cost/service performance must be based on effective management of complex transaction processing, including documentation preparation and customs clearance; and the effective use of third parties, such as customs agents, freight forwarders, banks, and so forth, takes on added importance.

It remains unclear how broadly the concept of global logistics will evolve. Clearly, many companies will never adopt a global network. However, after years of reading about the global corporation and the global logistics concept in the business press, I see an increasing number of companies actively establishing global logistics systems. Recognizing this, forward-looking logistics managers are expanding their skill base to prepare themselves for their new responsibilities.

7.19 The Realities of European Integration

October 1992

When formalized a number of years ago, the European Free Trade Pact initiated a period of negotiation and agreement on over 300 initiatives designed to unify a market of 12 countries and over 320 million consumers. As we approach the December 31, 1992 milestone for completion, much progress has been made. Scores of initiatives have been finalized, cross-border barriers have been greatly diminished, an agreement on monetary unification has been presented to each country for consideration, and many companies have begun rationalizing their European operations.

From a historical perspective of 10 or 20 years, it appears that much has been accomplished. One would expect that proponents of a unified Europe would be pleasantly surprised by the extent of these developments. Yet the movement toward unification must be disappointing to its sponsors, who envisioned considerably more progress toward integration by this time.

This gradual move toward integration—with its many stops and starts and resultant uncertainty—presents a challenge to logistics managers. I find many logistics managers in a quandary over how to proceed with planned changes to their European distribution systems. There is no simple answer I can offer; solutions differ considerably depending on the industry sector. But an understanding of the key forces driving these changes in Europe will provide guidance for the logistics manager operating in this complex environment.

- **Product standardization**—Two forces have slowed the pace of product rationalization and standardization for many companies. One is the slower than expected development and implementation of Europe-wide technical specifications and standards. Progress in this area varies by industrial sector, but the future direction is clear. Standards will emerge over time.

  The second force is a more challenging one for logistics managers. National tastes and preferences have limited the opportunities for product rationalization and therefore also limit a company's ability to leverage a centralized inventory. In Germany, for example, large aspirin tablets are preferred. French consumers, on the other hand, prefer smaller pills. These distinct geographic market requirements may preempt a centralized stocking location, which would reduce a company's inventory investment and total logistics costs.

■ ***Congestion***—Severe traffic congestion lengthens transportation lead times and makes companies uncertain about their ability to meet delivery schedules. This problem is not likely to lessen within the next decade, if at all, and it limits the feasibility of a consolidated distribution system. It is an important issue to which U.S. and Japanese companies in particular often do not give proper consideration.

■ ***Non-structural barriers***—Probably the greatest inhibitors to a rationalized manufacturing and distribution structure in Europe are non-structural barriers such as national cultures, business organization, and information systems capabilities. These national characteristics make centralized manufacturing and distribution more challenging to implement. Moreover, most European companies have traditionally been organized with managing directors heading operations in each country. Centralized manufacturing and distribution takes authority and responsibility away from these powerful managers. As a result, even though such changes offer dramatic cost benefits, many companies have had difficulty implementing new organizational structures.

Moreover, many companies have found their progress toward a rationalized, more centralized operations structure has been restricted or delayed by the need to change both their operating procedures and the supporting information systems. They are finding that they must totally rethink their forecasting, production planning, inventory planning, and distribution planning processes. In many cases, they have had to integrate incompatible, country-based information systems to allow Europe-wide operations.

There is no question that full European economic integration will occur. You should expect, however, that it will evolve over the next two decades and that each industry segment and even individual companies will experience differing rates of progress. This slow evolution poses challenges for logistics managers as they try to develop interim operating strategies. One thing is clear, however: the key to logistics strategic planning for Europe will be flexibility.

7.20 The Narcissistically Dysfunctional Organization

August 1996

Real leadership can truly inspire. A manager who can create a sense of community and a sense of common purpose among his or her team and then develop an environment which inspires each individual to excel and to feel genuinely empowered can create a powerful, high-performing organization. Unfortunately, too many organizations are run in a different way. Many managers are consumed with their self-importance. They create an environment which is rigid, not open to new ideas and constructive dialogue, and operated in a top-down, controlling fashion. These organizations (departments, divisions, companies, etc.) invariably underperform and do not reach their potential. And amazingly, top management often does not even notice.

The worst of these organizations is what I call the *narcissistically dysfunctional organization*. Narcissistic is derived from Narcissus, a youth in Greek mythology who was so self-consumed that he constantly sat by a reflecting pool admiring his own image (by the way, he was eventually transformed into a flower). Dysfunctional simply means operates in a non-effective, even destructive, way. The narcissistically dysfunctional organization is a miserable place to work and a poor performer. The symptoms of a narcissistically dysfunctional organization are easy to spot:

- The leader of this organization puts a greater premium on his or her time than anyone else's. He or she keeps people waiting for a meeting, takes telephone calls while in a meeting with subordinates, and shows a disrespect for their time. In a healthy organization, the leader understands that everyone's time is valuable and operates accordingly.

- In a misguided attempt to "lead," the manager of the narcissistically dysfunctional organization sets policy and makes decisions without counsel or consent from the team. A true leader seeks ideas, input, and counsel from a wide range of people (who see things that the manager does not); attempts to understand their ideas and perspectives; and works to forge a consensus among the team. This approach results in better decisions because more information and the best thinking of everyone in the organization are considered in the decision process.

- When was the last time your boss asked you, "How are we doing? How can we improve our performance or our working environment?" If you answer is "not recently," you might work in a narcissistically dysfunctional organization. A true leader and effective manager constantly seeks input on how to improve overall performance or how to create a better working environment.

- The narcissistically dysfunctional manager will scold individuals or the team when individual or collective performance (sales, shipments, productivity, profitability) dips. This style often reflects an insecure, perhaps even sadistic, character. The true leader will share good and bad news with his or her team in a respectful way, with the aim of sharing in successes or collectively finding ways to improve performance. A scolding or intimidating approach is a poor way for a father or mother to manage a parent–child relationship; it is even worse when one brings this style into the workplace.

The narcissistically dysfunctional organization robs each employee of his or her dignity and severely limits how a company or department can perform. There is no place for this approach in today's business world. It is beyond time for many companies and individuals to move to a higher level.

INDEX